REVOLUTIONS AND RECONSTRUCTIONS IN THE PHILOSOPHY OF SCIENCE

REVOLUTIONS AND RECONSTRUCTIONS IN THE PHILOSOPHY OF SCIENCE

MARY HESSE

INDIANA UNIVERSITY PRESS
Bloomington and London

Manufactured in Great Britain

Library of Congress catalog card number: 80-7819
ISBN 0-253-33381-4

Contents

Abbreviations used in Notes and Bibliography

Am. J. Physics	American Journal of Physics
Am. Phil. Quart.	American Philosophical Quarterly
BJPS	British Journal for the Philosophy of Science
Rev. Met.	Review of Metaphysics
Proc. Arist. Soc.	Proceeds of the Aristotelian Society

Introduction

The recent revolution in empiricist philosophy of science may be characterized in many ways, not least important of which is the turn from logical models to historical models. The standard account of scientific explanation due to Carnap, Hempel, Nagel, Braithwaite and Popper not only presupposed the methods and tools and consequent rigour of mathematical logic, but also viewed the justification of science itself as an essentially logical enterprise. But the subsequent historically oriented work of Kuhn, Feyerabend and Toulmin, and the epistemology of Quine (partly deriving from another historian of science, Pierre Duhem), has now undermined several of the premisses upon which the standard empiricist account depended. The most important of these are the assumptions of *naïve realism*, of a *universal scientific language*, and of the *correspondence theory of truth*.

These three assumptions between them constitute a picture of science and the world somewhat as follows: there is an external world which can in principle be exhaustively described in scientific language. The scientist, as both observer and language-user, can capture the external facts of the world in propositions that are true if they correspond to the facts and false if they do not. Science is ideally a linguistic system in which true propositions are in one-to-one relation to facts, including facts that are not directly observed because they involve hidden entities or properties, or past events or far distant events. These hidden events are described in theories, and theories can be inferred from observation, that is, the hidden explanatory mechanism of the world can be discovered from what is open to observation. Man as scientist is regarded as standing apart from the world and able to experiment and theorize about it objectively and dispassionately.

Almost every assumption underlying this account has been subjected to damaging criticism. It was realized from an early stage in the scientific revolution that there is a logical problem about the inference of theories from observations, because this cannot be made logically conclusive: there are in principle always an indefinite number of theories that fit the observed facts more or less adequately. Since renewed attention has been given to this problem by Duhem and Quine, it has come to be called the *underdetermination* of theory by empirical data.[1]

Until recently it has been possible to regard this thesis as a mere logical oddity—an example of the fallacy of deducing general premisses from their singular consequences without the aid of further general premisses. It could be held that, although underdetermination is in principle possible, in practice theories that are constrained by increasing amounts of evidence in a given domain do tend to exhibit convergence upon generally accepted theoretical truth, because ultimately at most one of the possible logical alternatives can be found to satisfy other 'natural' conditions upon scientific theory, for example, sufficient consistency and simplicity to be logically manageable, and sufficient plausibility to accord with acceptable presuppositions about such general categories as space, time, matter and causality. I have called all such general conditions upon theories 'coherence conditions', to indicate that the acceptability of a theory is partly judged by its coherence with them, as well as by its 'correspondence' with empirical data.[2] So the present objection to the thesis of underdetermination can briefly be expressed in what we may call the *convergence formula* (C):

Accumulating data plus coherence conditions ultimately converge to true theory.

There are however, two sorts of arguments that indicate that this is an over-simple formula. Some of these are epistemological and others are historical. First there is a problem about how to express the data, which is raised by the thesis of *theory-ladenness*. Every 'observation statement' describing empirical data has to be expressed in some language containing general descriptive predicates. Every set of predicates in a natural or any other descriptive language implies a *classification* of the contents of the world. Aristotle believed that a natural classification of kinds or essences is given by 'intuition', but this view has turned out philosophically and scientifically untenable, not least because the intuitions of natural kinds claimed at various stages of scientific development have turned out scientifically false. Modern versions of essentialism therefore tend to rely on science itself for the discovery of natural kinds: natural kinds are those that conform to the best scientific classification and system of laws. For example, the true classification of the chemical elements awaited the delineation of chemical laws by Lavoisier and his successors, and this, it is claimed by essentialists, is continually being improved upon by better and better approximation to a true system of laws in later chemistry and physics.

But the original argument is now proceeding in a circle. We began with the accumulating data as one element in a convergent process leading to true theory. But we have now found that in order to express the data in the form of true propositions, we have to *presuppose* that they are expressed in a language based on true natural kinds, and true natural kinds can only be identified if we know what is the true theory. Our convergence formula may, of course, be interpretable as some form of successive approximation in which this circle is acceptable, but if that is the case the nature of this process requires further investigation, as we shall see below, and the formula certainly cannot be taken as the basis of any simple interpretation of what constitutes 'true' theory.

The second epistemological problem concerns the coherence conditions. Those who wish to take (C) as a justification of the notion of uniquely true theories, are obliged to regard the coherence conditions themselves as uniquely true. Now there are in the history of philosophy arguments that purport to show that certain principles such as those of logical consistency, simplicity, and certain categorial properties of space, time, matter and causality, are necessarily true of natural knowledge as such. Some of these arguments, such as Aristotle's, attempt to derive *necessary properties* of the world; others, such as Kant's, attempt to show that the principles are necessary for any claim to *knowledge* of the world. There are very severe objections to all these arguments, both philosophical and scientific. I refer to some of the philosophical objections in chapter 2, but, as in the case of arguments to natural kinds, the most telling objections are probably the scientific ones: every set of metaphysical or regulative principles that has been suggested as necessary for science in the past has either been violated by subsequent acceptable science, or the principles concerned are such that we can see how plausible developments in our science would in fact violate them in the future. As for principles of simplicity, there are so many different versions of these that their formulation has to be tailored to actual theories rather than theories to them; and as for the categorial principles, almost every traditional principle of space, time, matter and causality has been violated in modern physics.[3]

Philosophical arguments in these matters come and go, but it is the demonstration within the *history* of science that no simple convergence as suggested by formula (C) actually occurs that has revived the importance of underdetermination. In the wake of Kuhn's *Structure of Scientific Revolutions*, many studies have laid emphasis on the revolutionary conceptual changes that take place in the sequence of theories in a given domain of phenomena. The importance of historical examples is not only persuasive, but they also help to clarify the sense in which formula (C) is correct and the sense in which it is mistaken.

Consider the particular sequence of conceptual revolutions in the history of the chemical elements. In the early stages of this sequence, from the Aristotelian four elements, through the phlogiston theory, to Lavoisier's list of elements, it may look as though there is no convergence, for each mode of classification of matter disrupts and reorganizes the classification that went before. There is no obvious sense in which convergence of *concepts* can be maintained. After Lavoisier, however, through Dalton, Avagadro, and into modern physical chemistry, there is a form of convergence, in the sense that the list of elements originally proposed by Lavoisier has been modified and added to, but not radically reformulated. Each successor theory in this sequence interprets its predecessor in its own terms as an approximation to itself. This 'correspondence postulate', relating earlier to later theories, has even become a condition upon the development of advanced physical science.

It is this aspect of the development that accounts for the

fact that, in spite of the underdetermination thesis, natural science is *instrumentally progressive*. There is instrumental progress in the sense that we now have vastly increasing pragmatic possibilities of predicting and controlling empirical events by means of experimentation and theory construction. This progress takes place not only in the sense of a numerical approximation of one theory to the next, but also, as in the case of the list of elements, because the concepts of one theory are usable with sufficient approximation at certain levels of accuracy even in the context of the next, conceptually different, theory. But if the fundamental theory underlying chemistry from Lavoisier to quantum theory is taken to give the best-up-to-now description of what the elements actually *are* in each theory, then radical differences in the connotations of the concepts arise—from classical atomism to force-field theory to atomic nuclei plus electron orbits and most recently to quantum fields.

This is the sense in which instrumental progress does not entail convergence of the conceptual framework of theories towards universalizable truth. At best it entails increasing approximation of low-level laws and predictions to the subsequently tested data, and that this approximation takes place in limited domains of data. Pragmatic success and approximate laws are always relative to particular local phenomena. Even if these phenomena extend to the galaxies, they never encompass the possibly infinite universe, and however extensive they are in space, they are necessarily very local in comparison with the whole past and future of the universe in time. The whole of the data is never in, and there is always room for further conceptual revolution, however accurate the current theory is for current purposes.

Local approximation does not entail universalizability of conceptual frameworks. In fact the truth claims of a given stage of development are strictly speaking *non-propositional*, for expressions of what we know in one conceptual framework do not satisfy the axioms of propositional logic: there is no transitivity of entailment for sentences that are only approximately true, and no transitivity of application of general terms that are only justifiably applied within limited domains of phenomena. As another example, consider the way in which atomic electric charges gave way in Maxwell's work to the field interpretation of charges as the interruption by electric conductors of lines of force in free space or in electric insulators.[4] All the low-level laws of behaviour of electricity remain approximately the same, but the relation of the concepts involved is one of total reinterpretation. What was previously 'substantial' (the electric charges), becomes an epiphenomenal property of motion becomes energy of field of force), and energy of motion becomes energy of stress. The subsequent development of quantum theory carries another radical reinterpretation of electric charge. There is no convergence to an ideal conceptual language, and nothing suggests an end to the sequences of conceptual revolutions as long as theoretical science itself continues to develop.

Thus the thesis of underdetermination is shown to be more than a trivial logicality, and formula (C) is seen to be misleading just in respect of those implications of theories that have most extra-scientific interest, namely the picture of the world given by their conceptual frameworks. Here all pretensions to universalizability and metaphysical necessity in what we can know must be abandoned. No actual theory at any given stage of development can be said to be empirically true in a strictly propositional sense.[5] Hence most of the philosophically interesting conclusions that have been drawn at one time or another from current fundamental theory have no grounding in established scientific knowledge, for example, determinism or indeterminism, the finiteness or infinity of the world in space and time, the continuity or atomicity of matter in space and time.[6]

In view of the prestige that scientific theory has acquired as perhaps the only reliable and universalizable mode of knowledge, this has of course profound consequences for epistemology in general. One way of attempting to mitigate its consequences is to abandon positivist epistemology rather than the claims of theories, that is, to maintain the meaningfulness of ideal true theory, even though we cannot know it. This is the response that has occurred in the revival in Anglo-American philosophy of a metaphysical form of scientific *realism*, which effectively neglects epistemological questions in favour of analyzing the ontology of theories as if current science is known to have arrived at or somewhere near the truth.[7] A feature of this type of realism is that it retains the logical presuppositions of empiricism, namely the accurate applicability of logic and an ideal scientific language to the world. But if I am right in locating the problem of truth in the approximation and non-universalizability of our conceptual frameworks and hence of what we can know in science, this logical presupposition is groundless, and even if it is true, it is inapplicable. Logic is itself an ideal to which real theories never perfectly attain, and we should beware of drawing consequences of importance to the philosophy of real science from the presupposition of this ideal.

The revival of metaphysical realism after decades of positivism is one of the most remarkable features of current analytic philosophy. It may perhaps be traced to a fundamental difference of interest between analytic philosophers of the present generation and those who founded the positivist schools of epistemology. This difference may briefly be expressed by saying that the latter were concerned with what we can know to be true, and wished to repudiate metaphysical claims to knowledge, specially because these tended to be associated with Hegelian tradition not only in philosophy but also in authoritarian religion and politics. Liberalism and social democracy, the Vienna positivists thought, were better defended in association with naturalistic knowledge.[8] Modern analytic philosophers, on the other hand, have generally divorced their philosophical from their ideological and practical interests, and have therefore lost the urgency of the question 'How can we know?', being content to presuppose ideal ontologies to which they feel no need to resort for the answers to practical questions.

In a wider perspective this realist problematic looks parochial and over-intellectualized. It has not only underestimated the challenge to empiricist presuppositions arising from modern history and philosophy of science, but it also bypasses two other features of the general philosophical scene. The first is the hermeneutic understanding of the human sciences as developed in Continental tradition from Dilthey and Weber to Gadamer and Habermas, and also in

Anglo-American philosophy by the successors of Wittgenstein. The second is the sociological discussion of epistemology initiated by Marx and continued by Durkheim and their successors. These approaches, together with the more recent ones deriving from Quine, Kuhn and Feyerabend, have led to what many realist philosophers would regard as the threat of *relativism*. Indeed it is often unreasoning fear of relativism that goes far to explain the power with which metaphysics seems to have caught the imagination of the realist philosophers. The relativist view of science, generally speaking, accepts what I have said about its pragmatic elements, but tends to play these down in comparison with an emphasis on theory, which is yet consistent with recent philosophy of science. In extreme forms of relativism theories are regarded only as internally connected propositional systems, or 'language games'; they are world-views to be given significance in their own right. 'Truth' is defined as coherence with the theoretical system, and 'knowledge' becomes socially institutionalized belief. The view is 'relativist' in the sense that there are no cross-theory criteria for belief, nor progressive approximations to universally valid knowledge in the theoretical domain.

If we take the thesis of underdetermination of theories seriously, relativism is a consequence that is inescapable in some form. But it appears particularly damaging just insofar as the ideal of knowledge retains the empiricist presuppositions of naïve realism, the universal scientific language, and the correspondence theory of truth. Like instrumentalism, relativism abandons claims for theories which rest on these assumptions, but appears to put nothing adequate in their place. The papers that follow in this collection all address themselves in one way or other to a critique of the empiricist presuppositions, and try more positively to steer a course between the extremes of metaphysical realism and relativism. In the end, as I indicate in Part III, I do not believe that these problems will be resolved in the context of natural science alone, but that they will require integration of natural science into a wider epistemological framework embracing the philosophy of social science, hermeneutics, and the sociology of knowledge.

The historiography of science

The first two papers in this collection are concerned with the consequences of the philosophical revolution for historiography of science. Belief in natural science as the progressive discovery of naïvely realistic truth about the external world made the writing of history of science comparatively straightforward. The historian had only to tell the success story of discovery, with its consequent value judgments upon both the results and the methods of past science as either progressive or erroneous, that is, as steps towards the best science now known, or as deviations from it. Reinterpretation of the history of science in terms of sequences of conceptual revolutions has made this kind of inductivist historiography impossible. Not only has it called into question the claim that scientific theories approach closer and closer to a comprehensive description of the real world, but it has undermined inductivist judgments about what constituted good science in the past, and even questions the unique rationality of modern science as the road to knowledge. In 'Reasons and Evaluation in the History of Science', I consider the consequent problems for interpretation of past science. One alternative to inductivist historiography is to abstain from evaluations of past science altogether, and to attempt to understand it in terms of its own internal criteria and interrelations with the thought-forms of its age. This idealist approach, I argue, is equally unsatisfactory and indeed equally unhistorical. The writing of history is a relation between two periods—that written about and that written from. Some elements of inductivism and its consequent evaluations are bound to enter *our* view of past science, since what counts as past *science* is partly determined by what we perceive as its historical continuity with our own.

In a hitherto unpublished paper, 'The Strong Thesis of Sociology of Science', I take up a different set of historical problems arising from the underdetermination of theories by empirical data. Where the idealist approach to historiography sought to explain the character of theoretical science in terms of the conceptual framework, or metaphysics, of its age, a more recent approach has been the sociological one of seeking explanations in terms of more comprehensive social factors. According to the 'strong thesis', it is not only deviations from rationality in the history of science that require causal explanation in terms of external social factors, but also 'true' science and 'rational' method. The thesis therefore raises again the questions of what are 'truth' and 'rationality' with regard to science past and present, and also poses more difficult questions in the sociology of knowledge. For example, does the strong thesis imply an unacceptable relativity of rational categories with respect to local social and cultural norms? what kind of knowledge status can be claimed for the strong thesis itself? does the strong thesis imply social determinism, or the causal priority of substructural over ideological factors? These questions are discussed in the light of several recent examples of social orientation in the historiography of theoretical science.

Objectivity and truth

The second group of papers takes up directly some problems arising from post-empiricist interpretations of science. In 'Theory and Observation' I explore the consequences of replacing the deductivist account of scientific explanation by what I call (following Quine) the *network model*. Here no fundamental distinction is made between observational and theoretical languages. Both are assumed to have the aim of true description of the world, but the concept of 'true description' is not simply that of correspondence, but involves a more complex account of the application of learned predicates in particular empirical situations, together with the connection of these in scientific classifications and theories, and subsequent prediction, test, and confirmation or correction. Neither theory nor observation language are immune from correction; the distinctions between them concern rather their relative directness of empirical reference and application, and the relative entrenchment of terms in the natural and theoretical languages. In this model, science retains its empirical basis, because the initial criteria of learning the correct use of descriptive terms in the natural language are empirical, and the self-corrective feedback process depends essentially on recognition of the success or failure of empirical predictions. The account therefore retains also the essentials of the correspondence theory of truth, but without the assumption of a stable observation language unpermeated by theoretical interpretation. The view of truth is, however, also essentially *instrumental*, since it derives from situations of prediction and test, and its relation to theories is indirect. Since the thesis of underdetermination of theory by data is built into the model, the sense in which 'truth' can also be predicated of theoretical frameworks remains undetermined.

'The Explanatory Function of Metaphor' explores in more detail the use of language in relation to theories. Here the mutual give and take between novel theoretical concepts and natural observation language, which is implied by the network model, is described in terms of what Max Black has called the 'interaction' theory of metaphor. I suggest that in the sense of this interaction view, theoretical explanation can be interpreted as metaphoric redescription of the domain of the explanandum. The requirement that the explanandum be deducible from the explanans has in any case to be modified in the network model by taking account of the *replacement* of initial observational descriptions of the explanandum by descriptions in theoretical categories, and it is natural to express this as replacement of inadequate literal descriptions by more adequate metaphor. Moreover, the problem of the meaning of theoretical terms, and their connection with empirical data by means of correspondence rules, is reinterpreted as a problem of metaphoric meaning: how are theoretical terms meaningfully developed by metaphor and analogy drawn from natural descriptive language? Theory as metaphor is understood in this view to have reference and truth value in the domain of the explanandum, and hence to be falsifiable.[9] It is also said to be consistent with a realist view of theories, although I should now want to modify that conclusion of the paper by taking account of the possibility of a *multiplicity* of more or less adequate metaphoric redescriptions, parallel to the possible multiplicity of undetermined theories. There is no 'perfect' metaphor.

The next paper, 'Models of Theory-Change', uses the

concepts of the network model to locate and compare a number of alternative accounts of theory structure. I argue that deductivism, inductivism, conventionalism and pluralism can all be seen as limitations upon the general network model. All these views, except some versions of pluralism, retain the criterion of empirical test and self-corrective learning (the *pragmatic criterion*) as essential to science, and I conclude that this criterion can conveniently be taken to demarcate what we currently mean by 'natural science'. Where the more traditional views of theory structure differ is in their attitude to the 'coherence conditions' for theories, that is, those conditions which are not directly derived from empirical test but which constrain the types of theory that are permitted in 'good' science. Again the question of the truth and realism of theories arises, now in the context of the possible truth claims of the coherence conditions. But it is concluded that science itself justifies truth claims only in relation to the pragmatic criterion.

The same thesis is pursued in more detail in 'Truth and the Growth of Scientific Knowledge'. There I try to show how the pragmatic requirement for a science to exhibit instrumental growth can be used to interpret the truth claims of science in any period or any culture, in spite of radical differences of conceptual framework. The account depends on a principle of 'charitable interpretation' of alien science, and shows how not only history but also epistemology of science requires some such hermeneutic principle. I conclude, however, that history of science and philosophy of science are not the same thing: historians should concentrate on science in its historical context, using modern analyses of the structure of science only to provide frameworks of interpretation; while philosophers must use some criterion of judgment derived from modern science and philosophy to construct 'ideal types' of science that are relatively independent of historical cases. Philosophers cannot forget that there is a history of *philosophy*—of the problems of empiricism and rationalism, realism and idealism—as well as a history of science.

A word of explanation is needed here about the status of the pragmatic criterion. I understand the criterion as the overriding requirement for empirical science to exhibit increasingly successful prediction and hence the possibility of instrumental control of the external world. I have effectively used it as the primary demarcating criterion between what counts as empirical science and what does not. There have, however, been many recent objections to the idea that there is any sharp demarcation of this kind. Historians and contemporary scientists are always able to point to cases of acceptable science that violate any criterion that may be suggested. For example, it does not appear that mathematical cosmology yields instrumental control, and on the other hand merely rule of thumb success in prediction does not count as science proper. All this is true, but to take it as conclusive against any demarcation of science is to misunderstand the relation between the historian and the philosopher. The philosopher is primarily interested in science as *knowledge*, and therefore in the kind of knowledge science claims, and in the conditions for its truth. Science claims empirical knowledge—if it does not, then there is nothing to distinguish it from any type of symbolic or metaphysical cosmology such as are pervasive in one way or other in every society. It is necessary to construct an 'ideal type' of what empirical knowledge is, and I take it as the lesson of recent discussion in philosophy of science that what empirical knowledge is reduces to the pragmatic criterion, together with, as perhaps a kind of bonus, the hope that successful pragmatic knowledge is best acquired through the search for comprehensive theoretical systems rather than rule-of-thumb instrumentalism. (That there is no guarantee of this latter point is the core of the problem of induction.)

Taking the pragmatic criterion as overriding is a philosopher's *reconstruction* of what now counts as science. It must fit existing practice fairly closely, but that does not imply that everything scientists do is directly relevant to it, nor that there are no other motivations for the development of scientific theory than the pragmatic. The criterion is the acceptable deposit of several centuries of empiricist understanding of science, and to take it as such no doubt involves a value judgment about the significance of the instrumental in our social context. No philosophical argument can ensure that the pragmatic criterion grasps the 'essence' of science: there are no such essences, only ideal types based on selective judgment. The criterion does, however, leave unanswered further questions about the significance of scientific theory, and these lie outside the limits of empiricism.

Some philosophers have responded to these questions by looking for new forms of scientific 'rationality' in the internal presuppositions and methods of what now and in the past has counted as acceptable science, divorced both from the pragmatic criterion and from social context.[10] But I have already suggested in relation to the 'strong programme' that it is not alternative internal rationalities that are required in the aftermath of empiricism, but rather a wider perspective on scientific theory in its social and ideological content. The question of rationality then becomes a question of the correct understanding of the place of science in society, rather than an idealist search for an autonomous superstructure. The validation of scientific theory, subject always to the overriding pragmatic criterion, is then seen in Durkheimian terms as the social legitimation of metaphysics and ideology, rather than as a quest for the grounds of empirical knowledge.

Pragmatic and evaluative knowledge

It is this shift of perspective on the theories of natural science that has provided one of the motives for my subsequent studies of the logic and methodology of the human sciences. 'In Defence of Objectivity' explores some alleged distinctions between the natural and the human sciences, as seen from the perspectives of empiricism and hermeneutics respectively. In empiricism, natural facts are said to be independent both of observer and of theoretical interpretation; natural laws are said to express external relations; scientific language is said to be exact, universalizable, and univocal; and the meaning of scientific concepts is independent of their application. Hermeneutic analysis of the human sciences on the other hand makes human meaning and intentionality constitutive of facts, regards theoretical interpretation as holistic and as internally related to these facts, and recognizes that the language of interpretation is metaphorical and dynamically adaptable to particulars. But in this list of distinctions, every one of the alleged characteristics of natural science has been questioned in post-empiricist philosophy. It is therefore possible that continuum rather than dichotomy is the more appropriate model for the relation of the natural and human sciences.

I discuss this possibility in terms of the differing *interests* of the two kinds of science as described in Habermas' *Knowledge and Human Interests*, and conclude by distinguishing two kinds of objectivity within both natural and human science:

1. Technical interest in external instrumental control, whose objectivity is ensured by the method of self-corrective learning. This interest applies to the more predictive aspects of the human sciences as well as to natural science proper.
2. Communicative interest, which includes the interpretive understanding of the human sciences, and also the social function of theoretical science, not as pure ontology, but as a mediation of man's views of himself in relation to nature. (This aspect of theory is developed in the work on sociology of science that I refer to in Chapter 2.)

In 'Theory and Value in the Social Sciences'[8] I develop the idea of a continuum of the sciences further by exploring the role of value judgments in constraining otherwise underdetermined social theories. I suggest that the crucial distinction between the social and natural sciences is not so much the presence or absence of evaluative ideologies, but rather the success or otherwise of the pragmatic criterion. In the natural sciences, this criterion is overriding, and enables ideologies to be filtered out in the historical development of a science, leaving a deposit of pragmatic or instrumental truth. There is no *a priori* guarantee, however, that the pragmatic criterion will be as successful in the social sciences, in other words there is no guarantee that these sciences will, can, or even should attain comprehensive and progressive theories like those of physics or biology. This fact, together with the admittedly evaluative character of adoption of the pragmatic criterion in the first place, suggests that the social sciences may properly adopt goals other than that of successful prediction and control of their domain. One alternative goal is Habermas' 'communicative interest', and the possibility of this and other goals is confirmed by various examples which unquestionably belong to the social sciences.

If new models of scientific rationality are wanted, it is in an analysis of the goals of knowledge rather than the internal techniques of natural science that they should be sought. Since the publication of *Knowledge and Human Interests*, Habermas has developed a more comprehensive theory of modes of knowledge and communication, and has moved away from the instrumentalism that seemed to be implied in his earlier work, and which I imputed to him in my 'In Defence of Objectivity'. The next paper, 'Habermas' Consensus Theory of Truth', attempts an exposition of his later theory, with particular reference to the problems of objectivity, truth and evaluation in natural science. This paper interprets Habermas' theory of truth as a shift from correspondence to consensus, with regard to both empirical and hermeneutic science. I examine the criteria of consensus in the so-called 'ideal speech situation', and argue, against Habermas, that these criteria are not unavoidable 'transcendental conditions of knowledge as such', but involve value judgments without ultimate rational grounds.

This paper must be regarded as no more than an interim report of one English reader's reactions to Habermas' prolific output. Since writing it, I have had the opportunity of reading his 'What is Universal Pragmatics?', first published in 1976,[11] where it becomes clear that Habermas has retained more features of correspondence theory in his philosophy of ordinary language and of natural science than I allowed for. I still think, however, that his writings do contain very fruitful suggestions towards a 'hermeneutic' analysis of the *theoretical* aspects of natural science.

Science and religion

In 'Criteria of Truth in Science and Theology' I consider some of the consequences of the foregoing view of science for the perennial science/religion debate. Like most philosophical debates that have presupposed empiricist analyses of science, this one takes on a different complexion when empiricist presuppositions are rejected. In particular, since naïve realism of theories is now abandoned, there cannot be any straightforward conflict between accounts of the external world as given in science and in theology. Disputes in the past that have depended on the realism of such accounts are not only based on misunderstanding of science, but also, I argue, of the nature of theological knowledge itself. The question arises, what *is* the status of theological claims to 'knowledge' of the world in, for example, doctrines of creation, of divine providence in history, and of the proper understanding of man? I conclude that these claims are *ideological* in the same sense that comprehensive theories in the human sciences are ideological, that is that they incorporate evaluations of their subject matter. Evaluations may be constrained by facts, but cannot be determined by them.[12]

In this sense theological claims do, indeed, rejoin the same category as scientific cosmologies, when these are themselves understood as the framework of social communication between individuals, groups, and nature. Thus science/religion debates can be re-engaged at a different level. This can indeed be seen happening in many popular expositions of scientific theory, which always go beyond truth claims justified by 'technical interest' to judge and evaluate the significance of nature to man and man to nature. It also happens in such extrapolations of currently accepted theory as that, for example, of Jaques Monod, who uses the results of molecular biology and theories of the evolution of matter to justify a universal ethics and metaphysics of 'chance and necessity'.

If 'scientific cosmology' is taken in the wider sense of any symbolic system situating human society in the natural world, then it becomes clear that the 'communicative' function here ascribed to scientific theory is just a continuation of the function of cosmology in practically every society, primitive and advanced. Cosmologies, scientific and religious, are Durkheimian 'collective representations', not closely tied to survival values as in narrower theory of functionalism, but relatively autonomous social creations which interact in many complex ways with the social substructure.[13] But here the threat of relativism becomes very pressing. Neither scientific realists, nor moral absolutists, nor religious believers, will easily tolerate the suggestion that theories, values, and doctrines owe their significance primarily to contingent social forms, however widespread in history or geography these may be. The cosmological beliefs of Judaeo-Christianity span millennia, and Western secular humanism, if not dominant, is now at least worldwide. But on a relativist view none of the theories, values, and doctrines embodied in these cosmologies are universalizable, much less necessarily true, and therefore fall short of the conditions that have always been held to be required of knowledge.

It may be, however, that this is where the usual discussions of relativism need to be stood on their head. It is common in liberal humanism to accept with alacrity the relativity of religious and ideologies to social context, to accept with somewhat greater hesitation and reluctance the like relativity of moral systems, but to baulk utterly at the relativity of scientific knowledge. I have argued that there are no defences except the pragmatic criterion against the relativity of scientific theories, and that the pragmatic criterion permits a permanent plurality of conceptual frameworks. I have also argued that no important philosophical or scientific positions are given away if we abandon the ungrounded assumption that there is an ideal, true conceptual framework for scientific theory. Ideologies are not created by the pragmatic criterion, but by evaluations, and apart from that criterion scientific theories share the characteristics of ideologies. Nothing is lost epistemologically if theories are taken to be relative to social context.

The same does not apply to ideologies themselves, however, nor to the moral systems they incorporate. The problem of their validity and their relativity is wholly different from that of scientific theories. Once they are distinguished both from the empiricism of the pragmatic criterion and from empiricist interpretations of scientific theory, the way is open to consider them in terms of modes of knowledge different from that of science. Hermeneutic knowledge is one such alternative mode; traditional systems of symbolism may be another. In these cases, unlike that of scientific theory, there is every reason to resist the relativist conclusion. The strength of the realist view of science lies in its empirical test procedure, and therefore the relativity of conceptual frameworks can be contrasted with the *real* attainment of *approximate* truth in pragmatic contexts. But in the case of postulated moral or ideological absolutes there is no empirical test procedure with which to contrast the appropriate mode of knowledge. Contingently speaking, relativity and plurality are facts. But it does not follow, any more than it does from the pluralism of scientific theory, that there is no non-relative truth in these domains. In science the assumption that there is is an idle assumption with respect to the pragmatic aspects of science, but in ideological contexts it is far from idle, since commitment to an ideology has immediate practical relevance in individual and social life. In science, historically relative conceptual frameworks are underdetermined by the pragmatic criterion, and exhibit no convergence or any other discernible relation to any ideally true ontology. For the pragmatic purposes of science it is not necessary that they should. But in ideological contexts, practical decision-making does depend on commitment to the content of actual conceptual frameworks, historically relative though these may appear to be.

The problem of how to validate ideology remains. Its conditions are left almost unchanged by the debate between realism and relativism in science, except that it can no longer be prejudged by any monopolistic cognitive claims for scientific cosmology.

Notes

1 Particularly in P. Duhem, *The Aim and Structure of Physical Theory*, English edn (Princeton N.J., 1954), Part II, chs 6, 7; W. v. O. Quine, *From a Logical Point of View*, Cambridge Mass., 1953, chs 1, 2; *Word and Object*, New York, 1960, chs 1, 2; *Ontological Relativity*, New York, 1969, chs 2, 3.

2 See my *The Structure of Scientific Inference*, London, 1974, p. 51 ff.

3 For a useful survey of quantum ontology see C. A. Hooker, 'The nature of quantum mechanical reality', *University of Pittsburgh Series in the Philosphy of Science*, vol. v, ed. R. G. Colodny, Pittsburgh, 1972, p. 67.

4 I have considered this case in *The Structure of Scientific Inference*, ch. 11.

5 I have previously formulated this in terms of a confirmation theory by means of the principle that all theories universally quantified in

potentially infinite domains have zero probability, where probability is interpreted as reasonable degree of belief (*ibid.* ch. 8).

6 These three happen to form the subject-matter of Kant's first three antinomies, about which he rightly said that neither alternative of each pair has empirical truth value. But there are others he did not recognize, for instance, the Euclidean or non-Euclidean character of space, the continuity or atomicity of space and time themselves, and also, probably, the question of the reducibility or emergence of the properties of life and mind from the physics of the inorganic. Cf. *The Critique of Pure Reason, Transcendental Dialectic*, Book II, ch. 2.

7 A balanced account is given in H. Putnam, 'What is 'realism'?', *Proc. Arist. Soc.* vol. lxxvi (1975/6), p. 177.

8 Cf. R. S. Cohen, 'Neurath, Otto', *Encyclopedia of Philosophy*, ed. P. Edwards, New York, 1967, vol. v, p. 477. It is remarkable how little of the political aspect of the thought of the Vienna Circle was mentioned in connection with its philosophical history until recently; compare for instance the articles in *The Legacy of Logical Positivism*, ed. P. Achinstein and S. F. Barker, Baltimore, 1969, with A. Janik and S. Toulmin, *Wittgenstein's Vienna*, London, 1973, especially ch. 8.

9 The account of theories as analogical, and their consequent falsifiability and confirmability, is developed in my *Models and Analogies in Science*, London, 1963, and *Structure of Scientific Inference*, especially ch. 9.

10 The proceedings of the Leonard Conference, *Scientific Discovery*, ed. T. Nickles, Dordrecht, 1980, contains many examples of work on this programme. The approach largely owes its inspiration to Popper, Kuhn, and Feyerabend, although these philosophers should not be held responsible for all the varieties of subsequent development.

11 English translation in *Communication and the Evolution of Society*, Boston, 1979, ch. 1.

12 I am pursuing these themes in the three annual series of Stanton Lectures in Cambridge, 1978–80.

13 Cf. E. Durkheim, *The Elementary Forms of the Religious Life*, English edn, London 1915, pp. 423–4.

I THE HISTORIOGRAPHY OF SCIENCE

1 Reasons and Evaluation in the History of Science

> . . . our proper conclusion seems to me to be that the conceptual framework of Chinese associative or coordinative thinking was essentially something different from that of European causal and 'legal' or nomothetic thinking. That it did not give rise to 17th-century theoretical science is no justification for calling it primitive. Joseph Needham, *Science and Civilisation in China*, II, p. 286.

I

The historiography of science, more than the history of other aspects of human thought, is peculiarly subject to philosophic fashion. This shows itself in two ways: first in the way historical studies reflect views of the nature of science current in contemporary philosophy of science, and second in the philosophy of history presupposed. An analysis of the first kind of influence was carried out by Joseph Agassi[1] in showing how inductivism and conventionalism generate distinctive kinds of history of science. The second kind of influence is exemplified by the application of Butterfield's category of 'Whiggish history'[2] to the kind of history of science that sees science as essentially cumulative and progressive, or by Collingwood's use of history of science in *The Idea of Nature*[3] as a paradigm case of doing history according to the Hegelian prescription of 'thinking men's thoughts after them'.

Historians of science have been on the whole less self-conscious about such issues than their colleagues in general history, and philosophers of history have devoted little attention to the special historiographical problems of science. In the last few decades, however, history of science has come of age as a sub-discipline of general history, and consequently some of these problems have begun to be discussed, not least in the sensitive comparisons of methodology to be found in Joseph Needham's work from which I have quoted above. My aim in this paper is to bring out into the open one such problem, which can best be described as the question of whether *evaluations* of the truth and rationality of past science are proper parts of the historian's task.[4]

That this is only now becoming a problem in the history of science may well come as a surprise to general historians, who have long wrestled with the problem of evaluation in other fields, and arrived at some implicit understanding that a kind of evaluation of the past is a necessary condition of good history. As W. H. Dray puts it:

> Historians . . . who concentrate largely on showing us how things appeared to the participants, it might be noted, are seldom regarded by their fellows as attaining the highest rank. They seem to be regarded, indeed, in much the same way that theoretical scientists regard those in *their* field who do not go beyond the level of classification and empirical generalization. In both cases, although in different ways, the idea of the inquiry requires a further interpretation of the materials thus provided.[5]

That the problem of evaluation is on the other hand a problem at all may come as a surprise to philosophers of science who instinctively regard any historical enquiry into science as incomplete which does not pose and answer the questions 'Was it reasonable?', 'Was it true?', and who in their more reconstructive moods are sometimes justly accused of preferring these questions to a more pedestrian investigation of 'Did it happen?'

There are, I think, three reasons why the best contemporary historians of science are not in a position to assimilate with simple piety the convergent wisdom of the historians and philosophers that evaluation of the past is both possible and necessary. The first is a form of backlash against the naïveties, both historical and philosophical, of the type of history of science which Agassi has called *inductivism*. This presupposed a philosophy of science according to which nature, when investigated with a properly open mind, reveals an ever-increasing accumulation of hard facts, which can be progressively better understood by means of cautious and tentative generalizations and modest theories. The process of scientific theorizing is in this view dangerous and has usually proved mistaken, therefore what is interesting in past science is the sum of its facts which still form the basis of modern science. The present is in this crude sense the standard of truth and rationality for the past, and gives the inductivist historian grounds for the reconstruction of past arguments according to an acceptably inductive structure, and for judging past theories as simply false and often ridiculous. Such inductive history is of course, among its other defects, self-defeating, because if all theories are dangerous and likely to be superseded, so are the present theories in terms of which the inductivist judges the past.

Replacement of inductivism by more sophisticated history and philosophy of science does not however entail that such history must avoid all judgments of truth or rationality. Inductivist judgments of the truth of past science as seen from the present may be anachronistic, but it has still been thought possible to evaluate the scientific character and reasonableness of past scientific inference by taking account of its more limited access to facts and different general presuppositions. Most of the best work in recent so-called 'internal' history of science has been conducted according to this recipe, which, again after Agassi, may be called *conventionalist*. Duhem rediscovered the continuity of astronomy and mechanics with the rational philosophy of Greece, Koyré and Burtt reconstructed its internal coherence in the period before Newton, Lovejoy described the 'archaic' background of biological science before the nineteenth century in terms of the 'great chain of being'. The tacit assumption of all this work has been that certain canons of rationality can be recognized even in long out-moded conceptual sytems, and that the injunction to 'think men's thoughts after them' can be followed as easily in history of science as in history of philosophy where it seems most at home, for natural science is just the arena of man's rational commerce with the world. It has also been presupposed in practice that this kind of internal history of science is relatively autonomous, that is to say that it can be carried on for the most part independently of the social and political environment of science and of the biographies of scientific researchers. But even if the autonomy of internal intellectual factors could be sustained, the tendency of internal history would still be towards relativizing canons of rationality along with the scientific ideas themselves, for an investigation such as Evans-Pritchard's into the metaphysics of witchcraft among the Azande[6] is not very different methodologically from, say, a study of Stoic physics.

However the autonomy of internal history with respect to environmental factors has not gone unchallenged. The second type of influence making for relativism in regard to history of science is the increasing application to it of categories of social history and psychological analysis, and resulting emphasis upon so-called 'external' or non-rational factors in the understanding of scientific development. I shall suggest later that the distinction between 'external' and 'internal' factors is by no means clear cut, since it has to accommodate a whole spectrum of influences from such social factors as the standards of education of a society, through unconscious psychological motivations, to the metaphysical commitments of an age and its accepted forms of scientific inference and logic. But there is no doubt that excitement over the comparatively new task of investigating the complex interactions between all these types of factor has challenged the independence of older types of internal history, and has tended to obscure the question of rational evaluation of scientific ideas and arguments.

A third development tending to relativize the historian's criteria of a rational science has been the failure of current philosophy of science to provide a generally acceptable account of scientific rationality which could serve to delimit the subject matter of internal history. If we consider the *deductivist* analysis of science which has been almost universally accepted by philosophers of science until recently, little help can be found for the internal historian seeking criteria of autonomy. The deductivist view has been characterized by a radical distinction between the sociology and psychology of science on the one hand, and its logic on the other, or as it is sometimes expressed, between the contexts of *discovery* and of *justification*. *How* a hypothesis is arrived at is not a question for philosophy of science, it is a matter for the individual or group psychology of scientists, or for historical investigation of external pressures upon science as a social phenomenon. The question for philosophy or logic is solely the question whether the hypotheses thus 'non-rationally' thrown up are viable in the light of facts, that is, whether they satisfy the formal conditions of confirmation and falsifiability adumbrated by deductivist philosophers, conditions which are themselves at present in a considerable state of disarray. Although this view places a heavy straitjacket on the philosophy of science, it appears to exert little restraint upon its history, and in particular it does not help historians to recognize timeless normative criteria of scientific argument. It allows historians to take seriously as scientific whatever theories were contemplated in the past, arrived at by whatever external or internal influences, and however apparently bizarre, just so long as these theories were, at least in intention, logically coherent and empirically testable. The use of the terms 'logical' and even 'rational' in the deductivist analysis is indeed far narrower than in the intellectual historian's 'internal logic of science', or in his view of the history of science as the history of man's rational thought about nature. For deductivism characterizes all influences leading to discovery as non-logical or even non-rational, and leaves the whole context of discovery to the efforts of the historian without offering him any criteria of distinction between kinds of influence on discovery. And since, for given evidence, a theory satisfying the deductive criteria is never unique, the particular kinds of concepts adopted are always dependent in this view on non-logical influences. Hence within deductivism as a view of science it is even impossible to make the distinction between intellectual and social influences on discovery, and *a fortiori* no general claim to internal autonomy of the history of scientific ideas can be sustained.

What has in fact happened is that, far from philosophy providing criteria for history, all forms of historical investigation, internal as well as external, have led to radical questioning of all received philosophical views of science. For they have revealed the impossibility of drawing any sharp line between the contexts of justification and discovery, between the 'rational' arguments as defined by deductivism and the psychological and cultural processes which determine what kinds of theory are contemplated, and even between the 'hard facts' which must be respected as tests of theory and the way these facts were interpreted in a given cultural environment. It is no accident that the current attacks upon these entrenched dichotomies of modern philosophy of science come either from historians of science (explicitly and recently from Kuhn, but also implicitly from Duhem), or from philosophers who are deeply immersed in history of science and conduct their discussions by means of detailed case histories (for example, Popper, Feyerabend, Hanson and Toulmin). Some of these writers hold a conventionalist position in philosophy of science which parallels what is described above as conventionalism in historiography, with stress on the role of intellectual factors in scientific development, but without any implication that external causes of change are excluded, or that there is any intrinsic distinction between the approaches of internal and external history.

II

How then are we to understand the evaluative task of the historian with regard to the rationality of the science of the past? The practice of historians is usually wiser than their theoretical self-reflections, and all these pressures towards scepticism about criteria of scientific rationality do not prevent the occurrence of evaluation and interpretation in the best historiography of science. But the pressures are not unimportant, because they sometimes tend to impoverish the self-understanding of historians of science, and sometimes to bias their practice and conclusions. In justification of these claims I shall consider in some detail a particular development in the historiography of seventeenth-century science. This is the effect of renewed interest in the hermetic and natural magic traditions, stimulated by Frances Yates' pioneering studies, particularly her book *Giordano Bruno and the Hermetic Tradition*, and by Walter Pagel's detailed work on Paracelsus and his period.[7]

The hermetic writings were a group of gnostic texts actually dating from the second and third centuries A.D., but believed in the sixteenth century to be contemporary with or prior to Moses, and originating in Egypt. They consequently carried all the ancient authority so much revered in the Renaissance; they were quite non-Aristotelian in spirit and hence reinforced the anti-scholastic tendencies of Renaissance thought; and since they were in fact written in the Christian era, they contained some elements of Judaism and Christianity which were regarded as prophetic and so enhanced still further their authority. P. M. Rattansi epitomizes well the main tenets of hermeticism in contrast with the careful distinctions maintained in medieval scholasticism between the natural and the marvellous, the magical, and the miraculous:

For Hermeticism, by contrast, man was a *magus* or operator who, by reaching back to a secret tradition of knowledge which gave a truer insight into the basic forces in the universe than the qualitative physics of Aristotle, could command these forces for human ends. Nature was linked by correspondencies, by secret ties of sympathy and antipathy, and by stellar influences; the pervasive nature of the Neo-Platonic World-Soul made everything, including matter, alive and sentient. Knowledge of these links laid the basis for a 'natural magical' control of nature. The techniques of manipulation were understood mainly in magical terms (incantations, amulets and images, music, numerologies).

Of this tradition Rattansi comments: 'It was not completely vanquished by the rise of the mechanical philosophy. Without taking full account of that tradition, it is impossible . . . to attain a full picture of the "new science".'[8]

In *Giordano Bruno* Miss Yates herself had specifically disclaimed the intention of contributing to the history of science proper: 'with the history of genuine science leading up to Galileo's mechanics this book has nothing whatever to do. That story belongs to the history of science proper . . . The phenomenon of Galileo derives from the continuous development in Middle Ages and Renaissance of the rational traditions of Greek science.' More recently, however, she has made bolder claims for the relevance of the hermetic tradition:

I would thus urge that the history of science in this period, instead of being read solely forwards for its premonitions of what was to come, should also be read backwards, seeking its connections with what had gone before. A history of science may emerge from such efforts which will be exaggerated and partly wrong. But then the history of science from the solely forward-looking point of view has also been exaggerated and partly wrong, misinterpreting the old thinkers by picking out from the context of their thought as a whole only what seems to point in the direction of modern developments. Only in the perhaps fairly distant future will a proper balance be established in which the two types of inquiry, both of which are essential, will each contribute their quota to a new assessment.[9]

The measured statements of Miss Yates and Rattansi only hint at the potential difficulties. Lying not far below their surface, and explicit in some younger writers, is the suggestion that the enterprise of internal history of science, as pursued for instance in the history of the seventeenth-century mathematico-mechanical tradition, is a mistake. It is a mistake because various kinds of non-rational factors are so closely bound up with rational argument that a history which tries to concentrate on the latter is necessarily distorted. It is a mistake also because, so it is implied, it reads and evaluates seventeenth-century science as part of a tradition which is our tradition, instead of understanding it in its own terms.

In order to investigate these issues we need to look more closely at the notions of 'rational science' and its 'internal history'. The way in which these terms are commonly intended by historians may be indicated in the hermetic example by contrast with some of the historiographical elements which are claimed to challenge them. These are (1) the social and political affiliations of certain religious sects, and the schools of Paracelsian and Helmontian doctors and chemists, (2) the full-scale hermetic and natural magic tradition as a way of thought and life in such writers as Paracelsus himself, Bruno, and Fludd, and (3) the doctrines of extended spirits and powers of matter which persisted in later seventeenth-century science, including Newton's work, in opposition to corpuscular mechanism.

Of the first of these factors it does not seem to be anywhere claimed that they provided more than the occasion and the motivation for certain developments connected with the new science. Such sectarian figures as Hartlib, Dury and Comenius helped to encourage Baconian allegiances in the early Royal Society, and the anti-establishment circles in which they moved go some way to explain the suspicion with which the Society was viewed by Royalists and Churchmen in the Restoration period. But none of this seems to impinge essentially on the internal tradition of the history of the mechanical philosophy as found in the writings of such historians as Duhem, Koyré, Burtt and Dijksterhuis. Indeed in the debates which have followed the related theses of Robert Merton and Christopher Hill regarding the influence of puritanism on seventeenth-century science, several commentators have remarked that the argument suffered from too little conceptual clarity about what was to count as 'science' (and indeed as 'puritanism'). Far from suggesting a restructuring of the internal tradition, these debates presupposed its existence, and the disputants were counselled to look at what had been achieved in internal history in order to acquire some internal specification of what 'science' is.[10]

The case with the second and third elements of the hermetic complex is different, because here it is not a question of interacting social factors, but of intellectual factors which might be held to be necessary ingredients of the history of science seen as the history of thought. Their close relation to the social factors just mentioned has obscured the fact that the real challenge to the received internal tradition comes not so much from social and political factors, as from within history seen as 'thinking men's thoughts after them'. In the passage quoted above Miss Yates excludes some *ideas* from the 'rational tradition', and yet these ideas are undoubtedly in men's heads and presupposed in much of their literature. The first question is, then, can the notion of 'internal history' be more closely defined in terms of some understanding of what is to be 'rational science', in contrast to complexes of ideas such as hermeticism? Only when this is answered can we judge whether the pursuit of relatively autonomous internal history is a viable proposition. If this question is taken in a philosophical sense it is a request for some perennial logical criterion which can be used to delimit the rationality of science at different periods, and it carries also the suggestion that a distinction of historical method between internal and external history is appropriate, since science pursued according to certain internal norms of rationality is likely to have a structure of development different from that of the contingent clash of events, actions and thoughts which are the stuff of general history.

In his book *Foundations of Historical Knowledge*, Morton White has given a useful and relevant analysis of the idea of intellectual history. He makes a threefold distinction between causal explanation, which may be rational or non-rational, and what he calls 'non-causal rational' explanation. 'Causal non-rational' explanations refer to the particular occasions upon which some belief comes to be held, for example the presence of certain Continental social reformers in England which encouraged some members of the earlier Royal Society to adopt Bacon's philosophy of science. On the other hand when a historian asks 'Why did Descartes believe in the existence of God?', he may give a 'non-causal rational' explanation in terms of 'the reasons stated by the thinkers or half-stated by them, or the reasons they would have stated if they had been asked certain questions'.[11] But, White goes on, it may be the case that 'whereas he *said* he believed p because it followed from certain other propositions which he thought were true, the *real* cause of his believing p was the fact that he wanted to believe it or was scared into believing it or was in the grip of some neurosis'. This kind of explanation White calls 'causal rational' explanation. Like all causal explanation in his view causal rational explanation falls under the regularity or covering-law model of historical explanation, whereas non-causal rational explanation does not. White appears to believe, further, that there can be no interaction between these different kinds of explanation, but at most a temperamental disagreement between different sorts of historians about what they want to call the *real* explanation. If this analysis were acceptable for history of science we could make the distinction between internal and external history coextensive with White's distinction between non-causal and causal explanation, and in terms of this definition internal history would be logically independent of all types of external factors.

Before discussing the adequacy of the analysis for history of science, it should be remarked in parenthesis that White's distinction is not the same as that commonly made in terms of explanation of human thought and action in terms of *causes* and in terms of *reasons*. Indeed his view of historical explanation is directly opposed to the view that history deals in reasons and not primarily in causes, that is that understanding of historical action (including belief) must involve a sort of sympathetic rehearsal of the intentions of the historical characters on the part of the historian, and that this method is logically distinct from the type of causal explanation appropriate in science. White's category of *causal* rational explanation on the other hand is intended to assimilate some beliefs that are commonly taken to be reasons (whether valid or mistaken) to causes as understood within the regularity view of explanation. It is only when reasons are appealed to as norms that they count as *non-causal* explanations; moreover appeal to norms must be by the *historian*, not only by the historical character, for the character's appeal might simply be reported by the historian as a fact to be inserted in a possible *causal* explanation. Thus it is only in connection with the history of intellectual pursuits, in which the existence of such atemporal norms is plausible, that White's category of non-causal rational explanation is applicable. Mathematics might be a fairly uncontentious candidate for such explanation, but philosophy and science are much more doubtfully so, as I shall now try to show.

If we attempt to apply White's distinction to the demarcation of internal history of science, we are immediately in difficulties. He appears to take for granted what constitute 'reasons' in the case of Descartes' argument for the existence of God. Are these, then, self-evident and timeless norms to which all philosophers *qua* philosophers are bound to conform? This can hardly be so, because even in cases where a classic philosopher can, as it were, be argued with as with a colleague, it is not self-evident that his own understanding and use of, for example, the ontological argument was the same as that of a modern analytic philosopher, and indeed some historians have argued that Descartes' understanding of this argument was not. And when it comes to the history of science, we have seen that neither science nor philosophy of science provide us with timeless norms of rationality ready-made.

Does White perhaps mean to assert that non-casual rational explanation must be given in terms of what were seen as reasons at the time? This is the kind of 'self-transcendence' or 'self-emptying' suggested, for example, by Lovejoy and Butterfield (both primarily historians, not philosophers), and by Collingwood. But White clearly regards this type of history as providing causal explanation, not reasons. A few pages before his threefold distinction he says:

> The savage's belief that a person is a witch may be causally explained by reference to another belief of the savage which, as *we* might say, does not logically support it. There are explanatory deductive arguments that connect one mad, false, or superstitious belief with others. And the point is that the historian is engaged in casual explanation *in the same sense* whether he is explaining the beliefs of a sane or insane man, a civilized man or a primitive man, a genius or a fool.[12]

The explanation is given in the same sense, not because all alleged logics are equally logic, but because the regularity model of explanation is the same for all types of cause, whether the causes are beliefs or not, and whether the beliefs are true or false.

It is clear that any proposal to try to make a logical distinction between atemporal rational explanation, and hence internal history, and causal explanation in terms of the rationality of a particular period will not stand up to a moment's investigation. In the sixteenth or seventeenth centuries there was no agreed 'scientific rationality of the time', and even the intellectual practice of individuals was a species of sleepwalking. What for example, in terms of this proposal, could we make of Copernicus' inference that the sun is at the centre of the planetary system because it is analogous to the king at the centre of his court, or to God the still centre of the universe? Indeed, as the historian has found when he has tried to specify seventeenth-century scientific method in its own terms, a very great variety of modes of acceptable argument emerge, some of them almost as remote from our views of rationality as the tenets of hermeticism itself. Moreover, metaphysics was inseparable from method: what was understood as 'science' was to some extent constituted by the mathematical, mechanical, non-animist, and non-teleological approach to nature. It is useless even to appeal to a general 'concern for facts' among those who count as 'scientists', for many of them are explicit that the mathematical and mechanical framework determines what is to count as a natural phenomenon, all else is excluded from science as supernatural or miraculous. Part of the historians' task is precisely to *discover* how far various kinds of inferences were acceptable at the time and why, and to investigate the *changing* conceptions of rationality which themselves partly constituted the scientific revolution.

The only remaining possibility of distinguishing internal history in terms of a logical demarcation of rational explanations is that reasons are just what appear to us to be reasons, whether or not we can explicitly formulate these, and whether or not there is any agreement about their timelessness or normative character. This might be called the *rational reconstruction* view of intellectual history, and it does indeed seem to be the standpoint adopted by White with respect to his non-causal explanations. It is, however, in conflict both with the judgments of historians quoted earlier, to the effect that autonomous internal history is necessarily distorted, and also with the actual tradition of internal history of, for example, seventeenth-century mechanical philosophy, where interpretations in terms of now-unacceptable forms of inference are taken for granted. Rational reconstructions, on the other hand, can be seen as a species of latterday inductivism in the history of science, in which, although past theories and alleged facts are not seen wholly in the light of their correspondence with present theories and facts, modes of scientific inference are so seen.

A case might be made for rational reconstruction along the following lines: there are sometimes in the development of science 'deep reasons' why one sort of theory is intrinsically preferable to another, or likely to be more progressive than another, and these may not be reasons that anyone was or could have been aware of at the time. By 'deep reasons' I mean logical or normative relations which emerge only if models of logically possible structures of science are investigated independently of their historical incarnations, and also mathematical relations which turn out to have interesting consequences which could not rationally have been foreseen at the time. As examples from normative philosophy of science we might refer to Popper's demonstration that a certain kind of power of theories is related logically to their simplicity, or to the proof within a probability theory of induction that inference by analogy from instance to instance gives high probability of predictions in some cases where theories understood merely as deductive systems do not. The first example 'explains' (in the sense of White's non-causal rational explanation) why scientists generally prefer simple theories, the second explains in the same sense why they usually proceed by analogical models rather than by formal deductive systems. The mathematical type of deep reason is perhaps applicable only to history of mathematics and the mathematically oriented sciences. An example would be Maxwell's introduction of the displacement current, for which he gave various relatively unconvincing reasons at the time, including an apparently superficial analogy with magnetism and also an argument depending on a dubious mechanical model of the aether. But though he could not have explicitly known it, the most significant feature of the displacement current was, and remains, that it renders Maxwell's equations Lorentz-invariant. It may not always be improper to 'explain' a scientist's apparently baseless intuition by such mathematical truths as this, particularly if our later knowledge of them enables us to interpret hints in Maxwell that could not have been intelligible to his contemporaries.

Passmore[13] has remarked that history of science differs from history of philosophy in that many histories written by philosophers are polemical, and are intended as first-order contributions to philosophical debate, but there are few polemical histories of science. Perhaps there ought to be more. Rational reconstructions may be a contribution to contemporary science, but their function in this respect should be distinguished from their contribution *to history*, where they should surely be seen only as ancillary to approaches which are more sensitive to different conceptions of rationality in the past.[14] Indeed no history of science can avoid mixing these ingredients in some proportion; the best history from Aristotle through Whewell and Duhem to Koyré has been such a judicious mixture, and has also been explicit about the normative philosophy of science adopted. To return to the question of the autonomy of internal history, it must be concluded that if 'internal' history is interpreted in any other way than as pure rational reconstruction, then the question whether it can be pursued independently of external and non-rational factors is a historical, not a philosophical, question, and the answer to it will vary from case to case. A little further consideration of the hermetic debate may illustrate this.

III

When all has been said about the absence of *a priori* criteria for 'rational science', our intuition remains that however varied may be the explicit and implicit methodologies of seventeenth-century science, they are still worlds away from hermeticism. This intuition is confirmed by several examples in the period of intellectual dispute between adherents of the two traditions in which we find moral, political, and theological arguments deployed along with philosophical ones in the course of vigorous repudiation of the hermetic cults.

In an exchange of polemics with the English Rosicrucian doctor Robert Fludd, Kepler dissociates himself from the interpretation of mathematics found in the hermetic writers.[15] Kepler does indeed himself believe in a mathematical harmony of the cosmos as the image or analogue of God and the soul, but his geometry is Euclidean, his conclusions require proof, and they must correspond with facts (that is, the kind of facts Kepler inherited in Brahe's planetary tables). According to Fludd, on the other hand, Kepler merely 'excogitates the exterior movement . . . I contemplate the internal and essential impulses'. Fludd *complains* that geometry is dominated by Euclid, while arithmetic is full of 'definitions, principles and discussions of theoretical operations, addition, subtraction, multiplication, division, golden numbers, fractions, square roots and the extraction of cubes'. There is, he goes on, no 'arcane arithmetic', no understanding of the significance of the number 4, deriving from the sacred name of God.

In less measured tones than Kepler, Mersenne devotes himself to combating the arrogance and impiety of the terrible magicians.[16] Their arbitrary numerologies do not even agree among themselves; they do not understand that words are mere *flatus voces*, merely conventional signs or sounds, not images or causes. The proportion of the planetary distances may exhibit harmony, but whether it does or not is a matter of fact, not of cosmic analogies. Moreover, astrology, magic and the cabala are not just harmless games, they reduce human freedom to cosmic determinism and hence are morally reprehensible. Although some alleged examples of sorcery may be facts, use of sorcery is morally detestable; the magicians are guilty of arrogance and impiety in their claim that the human intellect is divinely inspired and is the measure of things. When Fludd replies to this onslaught with equal violence. Mersenne requests Gassendi to take up the cause, and he, slightly reluctantly but for friendship's sake, drops what he is doing in order to study Fludd's writings.[17] That is a measure of the externality of the hermetics at this period to the new philosophy.

Another such polemical exchange is Seth Ward's *Vindiciae Academiarum*, written in reply to an attack upon the academic activities of the University of Oxford by John Webster.[18] Webster berates Oxford for its neglect of the new science, citing indifferently as representatives of that science Bacon, Copernicus, Galileo, Paracelsus, Boehme, Fludd, and the Rosy Cross. Ward replies with careful distinctions between the true natural language or universal character 'where every word were a definition and contained the nature of the thing', and 'that which the *Cabalists* and *Rosycrucians* have vainly sought for in the Hebrew'. Hieroglyphics and cryptography were invented for *concealment*, grammar and language for *explication*. Magic is a 'cheat and imposture . . . with the pretence of specificall vertues, and occult celestiall signatures and taking [credulous men] off from observation and experiment. . . . The discoveries of the symphonies of nature, and the rules of applying agent and materiall causes to produce effects, is the true naturall magic'. Both Mersenne and Ward take Aristotle for an ally against the magicians: it is not Aristotle, rational though wrongheaded, who is the enemy of the new philosophy, but 'the windy impostures of magic and astrology, of signatures and physiognomy'.[19]

Rattansi characterizes the situation accurately when he contrasts 'the emotionally charged and mystical flavour of Hermeticism, its rejection of corrupted reason and praise of "experience" (which means mystical illumination as well as manual operations), and its search for knowledge in arbitrary, scriptural interpretation', with a 'a sober and disenchanted system of natural knowledge, harmonized with traditional religion', and goes on: 'To move from one to the other was to change one conceptual scheme for ordering natural knowledge to another, with an accompanying shift in the choice of problems, methods, and explanatory models.'[20]

The change of sensibility is also a contemporary view. For example Glanvill: 'among the Egyptians and Arabians, the Paracelsians, and some other moderns, chemistry was very phantastic, unintelligible, and delusive, . . . the Royal Society have refined it from its dross, and made it honest, sober, and intelligible. . . .'[21] And Sprat's plea for a 'close, naked, natural way of speaking' is directed as much at the 'Egyptians' as at the Aristotelians.[22]

In view of all this, the suggestion of a *confluence* of hermeticism and mechanism into the melting pot of the new science is a mistake. In all that constituted its essence as a way of thought and life, hermeticism was not only vanquished by the mid-century, but had provided the occasion for the new philosophy to mark out its own relative independence of all such traditions. The style of argument required in the polemics is itself significantly different from that adopted in domestic scientific disputes. It involves rhetoric and ridicule, and appeals to theological and moral principle, and sometimes political and pragmatic test. A new form of rationality can be seen to be distinguishing itself from traditional modes of thought, this can be perceived and described by the historian, and in some cases its application can be seen historically to have delimited a relatively autonomous area of study. Thus the question about the autonomy of internal history is not a question of imposing external norms, but of investigating actual historical influences.

IV

In the light of the debate about hermeticism let us now look at three types of scepticism regarding rationality that have been drawn from the history of science. All three raise issues already familiar in the general philosophy of history, and I shall conclude by examining the question whether there are any peculiar features of natural science which make some assumptions of the general debate inapplicable to this case. The three sceptical conclusions are

1. Since no generally acceptable normative criteria of rationality are forthcoming either from philosophical or historical analysis of science, notions of what constitutes scientific rationality are historically relative.
2. Since we find in the history of science no guarantee of autonomous internal and external sectors but a complex of interactions between many different types of factor, to select some of these and omit others is necessarily to distort the picture.
3. Value judgments about the rationality and truth of past systems of scientific thought should be avoided by the historian.

1. I have already accepted that there are no *a priori* normative criteria for science suitable for providing a logical demarcation of internal and external history. It does not follow from this, however, that the relative autonomy of internal history, or of the history of a certain kind of scientific tradition, may not be established on historical grounds in particular cases. It is possible that it can be established in the hermetic example, and I would indeed judge that none of the literature on the hermetic tradition has yet shown the contrary. The possibility of finding within the seventeenth-century debates themselves claims to partial independence of one tradition from others; the difference in character of the argument within the mechanical philosophy and the polemics between it and other traditions; and already established results regarding the internal development of the physical sciences seen in terms of a relatively independent rational tradition implicitly defined by themselves, all suggest that relative autonomy may be justified on historical grounds. A proper historical perspective neither involves uncritical accumulation of every minor writing of forgotten figures, nor is it necessarily vitiated by the imposition of our standards of rationality on an alien age. We cannot, to be sure, merely adopt Butterfield's advice to use

the judgments of importance of the period rather than our judgments,[23] for how in that case could we judge the relative weight in the history of science to be given to Kepler's arguments as against Fludd's? To reply that the seventeenth century itself clearly accepted Kepler's and Mersenne's view rather than Fludd's is not sufficient, for this would be like relying on the popular verdict of Athens upon Socratic philosophy to dictate our judgments of intrinsic importance. So long as we select science as our subject-matter, we are bound to write forward-looking history in the limited sense that we regard as important what we recognize as our own rationality, having some historical continuity with our own science. This is likely to mean that given a choice of forms of explanation between what we regard as rational methods on the one hand, and social pressures or psychology on the other, we shall regard the former as more significant. But this does not imply that we impose our own theories or even our own views of method on the science of the past. And if it seems in danger of becoming a circular definition of internal history as that which is continuous with our science according to our internal history, the only cure is to look more closely at the historical record to see whether the relative autonomy of internal history can be maintained in spite of possible disturbing factors. We may, indeed, sometimes be shown wrong in our imputation of our sorts of reasons even when the conclusions were in our eyes correct, for we may find strong evidence that ideological or psychological motives were in fact more influential.

2. The problem of the distorting effects of selection is of course a perennial one for all types of historian, and might seem to be by now sufficiently well understood to make it unnecessary to give it special attention in relation to the history of science. But there have been so many unguarded comments on the hermetic tradition, and in other assessments of the relation of social history to science, that historians of science may almost be said to be unconscious of the problems. It often seems to be assumed that by adding to the picture all influences that fed into it, of all conceivable degrees of relevance, we get nearer to some form of complete

description or complete understanding of the 'whole picture'. But the view of history as complete description, or 'telling it like it was' is an error analogous to the error of inductivism in science. It presupposes that history is a search for hard facts, which are relatively independent of each other, and that the full picture is attained by accumulating as many of these as possible. As a philosophy of history, this is the temptation of the conventionalist rather than the inductivist, for the inductivist has already his principle of selection, whereas the conventionalist is likely to think that the total internal coherence of a period will emerge more clearly as more factors are taken into account. But even the suggestion that it is possible to get nearer the true picture by accumulating factors should be treated with caution. Throwing more light on a picture may distort what has already been seen. The immediate enthusiasm for searching for the names of 'Hermes', 'Orpheus', and other pseudo-priscine authors in seventeenth-century scientific writings may be more distorting than the received internal tradition itself, unless careful distinctions are made between several different types of use of these names. They may be used as pious archaisms; or as familiar labels for ideas about non-mechanical forces which are in fact central to the new scientific tradition; or as thinly disguised Christian piety, owing little to gnosticism and magic, as in the case of Kepler; or on the other hand as really indicating full-blooded adherence to a hermetic cult. Once all these distinctions have been made, the received internal tradition recovers something of its autonomy and is seen not to have been seriously distorted by their previous neglect.

3. Thirdly there is the question how far evaluations of the truth and rationality of past observations, theories, and forms of scientific inference are desirable or even unavoidable in the historiography of science. In the philosophy of the social sciences there are familiar arguments purporting to show that the selection of areas of interest to study, and even the occurrence of value-loaded concepts in these areas, does not imply that the investigator himself makes the value judgments. 'Hysterical' may be a value-loaded term in the history of medicine, but in this view the historian of

medicine can report the circumstances in which the term was used and with what implications, without himself pronouncing upon the desirability or otherwise of hysteria. This view should, I think, be kept distinct from the view that often accompanies it, namely that fact and *theory* can be kept distinct. It may be argued that the circumstances of application of 'hysteria' can be described without either making value judgments or using other value-loaded words (which would make the elimination of value impossible), but this does not entail that *theory*-loaded terms can be explained factually without appeal to other theory-loaded terms. Therefore rejection of the distinction of fact and theory (which I should myself want to reject) does not entail rejection of the distinction of fact and value.

Without pursuing that particular question further, let us assume that the consensus of analytic opinion is right, that in general the description by the historian of the use of value judgments by historical characters does not commit the historian to making these judgments, and in particular that various kinds of scientific inference and truth judgment can be described by the historian of science without any commitment on his part to the rationality of the inference or the truth of the judgment. It may, however, still be the case that there are peculiar features about the history of *science* which relate judgments of rationality more intimately to the subject matter than in the case of other types of intellectual history. I think there are such features, and I shall conclude by mentioning some of them to present a case for a limited inductivism in the history of science.

First it may be argued that the relativist interpretation of scientific methodology undermines the very assumptions of historical method itself. In the covering-law or regularity view of historical explanation, the deductive analysis of scientific explanation is explicitly taken as the model, and if the latter is abandoned there seems little case for retaining the former. The opposing view, associated with the ideas of irreducible human action, intention, and reason, on the other hand, has generally been defined in *contrast* to the notion of scientific explanation in terms of causes. As Nagel puts it, 'even extreme exponents of the sociology of knowledge admit that most conclusions asserted in mathematics and natural science are neutral to differences in social perspective of those asserting them, so that the genesis of these propositions is irrelevant to their validity'.[24] And even extreme exponents of the doctrine of *verstehen* as the mode of historical explanation require to use the results of so-called 'scientific history'—the provenance of documents, the dating of archaeological remains, the recovery of ancient languages, the changing ecology of environments, and so on. The modern practice of history would be impossible if it were not sometimes assumed to be meaningful to ask the question 'Did it happen?' in the expectation that generally speaking various kinds of historical investigation will converge upon the same answer. At least this expectation is similar to that generally made in science when an experiment is taken to yield an answer to the question 'What happens?' The two expectations have similar logical structures, even though it may not be correct to describe either of them in terms of a naïve inductivism. It is therefore viciously circular to use historical findings to undermine the timeless validity of these currently accepted forms of scientific practice.

In the particular case of the history of science this argument is even stronger. For if we are to explain, say, the fact that Priestley believed in the phlogiston theory, we have to consider what his data were, and what inferences he claimed to draw from them. But we know what his data were partly because he described them, and described them in terms such as 'I could not doubt but that the calx was actually imbibing something from the air; . . . it could be no other than that to which chemists had unanimously given the name of *phlogiston*'.[25] From this we have to recover *what* it was Priestley was observing, and this can only be done in the light of the best scientific information *we* have about what happens when lead oxide is heated in an atmosphere of hydrogen. Or take Oersted's twenty-year-long attempts to demonstrate what any schoolboy can now do with a battery, a piece of wire and a compass needle.[26] In order to understand Oersted's difficulties, we have not only to know that he believed in a Newtonian theory of central attractive and repulsive forces and applied it to produce the wrong expectations in this case;

we have also to reconstruct what his equipment must actually have been like in order not to have immediately revealed what is obvious to us about the direction of rotation of the compass needle. This is not an undesirable inductivism —it is a requirement of the programme of taking the facts and ideas of a period seriously. And it is incidentally one of the comparative tests of our science that it can not only explain what was differently explained in the past, but explain why other things were not explained or even observed when they should have been. It may of course be replied that this comparison is reciprocal, because *we* are doubtless also neglecting things we should be seeing, and a different rationality might detect and explain this neglect. But with regard to *our history* the point remains valid, for no one has yet suggested that *we* should write Aristotelian history of twentieth-century science, even if such a project were conceivable, and we certainly cannot write it from the point of view of a rationality of the future.

Examples from the history of science remind us that science has to do with happenings as well as ideas, and if the interpretation of past happenings in terms of modern theories is to be more than an arbitrary imposition of our own standards, it must be presupposed that there is something absolutely preferable in our own science to that of the past. That in other words there is a sense in which the development of science stands outside historical relativity and is absolutely progressive. Can this sense be perceived without losing sight of much that is acceptable in the relativist analysis?

Though the progressive character of science is not straightforwardly a question of having more knowledge about more facts, it is perhaps less misleading to put it that way than to refer to the progression of better and better theories or increasing adequacy of methods. What has been illuminating in the relativists' account is their demonstration that conceptual theories undergo deep revolutions and do not converge continuously towards some fundamental truth. For the same reason the progress of science is not a simple expansion of the number of true observation statements accounted for in successive theories, because the language of the observation statements is itself permeated by the theory, and there is no simple comparison between the way two theories describe the same events. Neither of course is it simply the case that we know more facts now than in, say, 1600. New facts may be learned by cataloguing, botanizing, collecting moon samples, or devotion to high-speed data processing systems. Some of these may be used by science and some may depend on science but they do not in themselves *constitute* the progressive character of science. Moreover post-1600 science has deliberately discarded and forgotten many facts that were then known—numerical facts about the Pyramids or Old Testament chronology, and facts about alchemical and magical operations for example. If science is in some sense progressive with respect to facts, the facts must be specified more closely as those that are significant for science.

There is however a sense in which we do know more facts, and that is the sense most closely related to technological control, although it is not simply identical with it. We know more facts that are accounted for and interpreted by very general and systematic theories, and we know them in the sense that we can use these theories to provide us with verifiable expectations about what will happen next. This is exploited in technology, but is not identical with it, because there could be rule-of-thumb technology without it, and on the other hand it need not be so exploited. Because this sense of the progress of science is about controlled happenings, it is independent of the way facts are described relative to different theories. It is a positive and absolute development of the last few centuries, and it constitutes the core of truth in inductivism, whether that is taken as a philosophy or a historiography of science. It provides an absolute criterion of distinction of our rationality, although it does not of course imply any *moral* evaluation of that rationality.

In summary, then, I have suggested three sources of scepticism in regard to the historian's treatment of scientific rationality: the collapse of inductivism, the failure of philosophical analysis to present a normative model for science which is generally acceptable and clearly relevant to the history of science, and the historical investigation of relations between history of science and social and ideological history.

I have argued that the notion of 'internal history' with which intellectual historians of science have worked is not definable in terms of external normative criteria of rationality, nor by simply taking the accepted criteria of the time as normative. Relative autonomy of internal history may, however, be justified in some cases, as with the mechanical versus the hermetic tradition in the seventeenth century, by historical investigation itself. The historian of science then inevitably adopts as his principle of significance that tradition which has some historical continuity with our own, although this does not mean that he imposes either our accepted theories or our accepted methods on the past in the fashion of the old inductivism or the newer rational reconstructions. Finally, I have suggested that our understanding of scientific rationality must have a special place in historical study, and especially where questions about what was actually observed have to be answered in the context of the history of scientific theories. In this sense, history of science is inevitably evaluative and forward-looking. Moreover this does not indicate a relativity to our science merely, because there is a relevant sense of 'progress' in which science is absolutely progressive in its understanding of facts. The sceptical conclusions lately drawn are not justified.

Notes

1 J. Agassi, 'Towards an historiography of science', *History and Theory*, vol. ii, London, 1963, 1.
2 H. Butterfield, *The Whig Interpretation of History*, London, 1931.
3 R. G. Collingwood, *The Idea of Nature*, Oxford, 1945.
4 For historical illustration I have drawn in this paper on an example developed at greater length in my 'Hermeticism and historiography' in *Minnesota Studies in the Philosophy of Science*, ed. R. Steuwer, vol. iv, Minnesota, 1971. I am grateful to the Minnesota University Press for permission to reproduce some paragraphs of that paper in the present one, principally in section III.
5 W. H. Dray, *Philosophy of History*, Englewood Cliffs, New Jersey, 1964, p. 38.
6 E. E. Evans-Pritchard, *Witchcraft, Oracles and Magic among the Azande*, Oxford, 1937.
7 F. A. Yates, *Giordano Bruno and the Hermetic Tradition*, London, 1964; W. Pagel, *Paracelsus*, Basle, 1958.
8 P. M. Rattansi, 'The intellectual origins of the Royal Society', *Notes and Records of the R.S.*, vol. xxiii, 1968, 131–2.

9 F. A. Yates, 'The hermetic tradition in Renaissance science' in *Art, Science, and History in the Renaissance,* ed. C. S. Singleton, Baltimore, 1968, p. 270.
10 R. K. Merton, 'Science in seventeenth-century England', *Osiris*, vol. iv, 1938, 360; C. Hill, *Intellectual Origins of the English Revolution*, Oxford, 1965; M. Purver, *The Royal Society: concept and creation*, Oxford, 1967; and articles by C. Hill and H. F. Kearney in *Past and Present*, 1964, 1965.
11 M. White, *Foundations of Historical Knowledge*, New York and London, 1965, p. 196.
12 *Ibid.* pp. 188–9.
13 J. Passmore, 'The idea of a history of philosophy', *History and Theory*, vol. v, 1965, 8, n. 18.
14 The distinction is made very sharply in J. Dorling, 'Scientist's history of science and historian's history of science', forthcoming.
15 The documents have been presented by W. Pauli, 'The influence of archetypal ideas on the scientific theories of Kepler' in C. G. Jung and W. Pauli, *The Interpretation of Nature and the Psyche*, English trans., London, 1955, p. 151. See also A. G. Debus 'Renaissance chemistry and the work of Robert Fludd', *Ambix*, vol. xiv, 1967, 42, and 'Mathematics and nature in the chemical texts of the Renaissance', *ibid.* vol. xv, 1968, 1.
16 R. Lenoble, *Mersenne ou la naissance du mechanisme*, Paris, 1943, p. 7; Yates, *Giordano Bruno*, ch. 22.
17 G. Sortais, *La Philosophie moderne depuis Bacon jusqu'à Leibniz,* vol. ii, Paris, 1922, p. 43.
18 S. Ward, *Vindiciae Academiarum*, Oxford, 1654; J. Webster, *The Examination of Academies*, London, 1654.
19 Ward, *op. cit.*, pp. 22, 34, 36; Lenoble, *op. cit.*, p. 146.
20 Rattansi, *op. cit.*, p. 139.
21 J. Glanvill, *Plus Ultra*, London, 1668, p. 12.
22 T. Sprat, *The History of the Royal Society of London*, London, 1667, section 20; see also section 3.
23 Butterfield, *op. cit.*, p. 24.
24 E. Nagel, *The Structure of Science*, New York and London, 1961, p. 500.
25 Quoted in S. Toulmin, 'Crucial experiments: Priestley and Lavoisier'. *Journal of Historical Ideas*, vol. xviii, 1957, 209.
26 See R. C. Stauffer, 'Speculation and experiment in the background of Oersted's discovery of electromagnetism', *Isis*, vol. xlviii, 1957, 33.

2 *The Strong Thesis of Sociology of Science*[1]

The strong thesis in the history of science

It is now a platitude to hold that the two approaches to the history of science labelled respectively 'internal' or 'rational', and 'external' or 'social' are complementary and not contradictory, and that any so-called conflicts between them are pseudo-conflicts. Both approaches have reached out from their original bases to embrace each other in a more genuinely and comprehensively historical stance. On the one hand studies like those of Yates, Pagel, Rattansi, Webster and Debus on the intellectual context of sixteenth- and seventeenth-century science have shown how the 'internal rationality' of science has to be widened to include the ideology of belief systems of all kinds—magic, mythology, alchemy, religious sectarianism—in other words all those cultural factors which in more Whiggish history were thought to be farthest removed from science.[2] On the other hand, 'external' or social history of science is no longer content only to study peripheral social factors in the provenance of science—for example educational and financial determinants of scientific work, or even the sociology and psychology of the scientific subculture in the manner of Kuhn. Social history of science is increasingly, and most interestingly, taken to mean study of the social conditioning of the theoretical belief systems of science—in other words sociology of science has become a branch of sociology of knowledge.

Among studies of this kind, some of the most notable are those of Paul Forman on quantum theory in the Weimar Republic,[3] papers in the Teich and Young volume *Changing Perspectives in the History of Science*, and some more recent case studies of which I shall give an account below. In addition to the particular case studies, three books have reflected upon the philosophical and historiographical issues involved: Barry Barnes' *Scientific Knowledge and Sociological Theory* (London, 1974) and his *Interests and the Growth of*

Knowledge (London, 1977), and David Bloor's *Knowledge and Social Imagery* (London, 1976).[4]

The sociology of knowledge is a notorious black spot for fatal accidents both sociological and philosophical. The theses connected with it are regarded by some as so clearly subversive of all good order and objectivity as to be beyond the pale of rational discussion, and by others as part and parcel of a variety of non-scientific commitments in ideology, morals and politics. In introducing a discussion of how the history of science can include the social history of scientific knowledge, I shall begin by considering some more general problems concerning the sociology of knowledge itself.

In his seminal work on sociology of knowledge, *Ideology and Utopia* (London, 1936), Karl Mannheim put forward two relevant theses. First he distinguished between *particular* and *total* theories of ideology in the following way. Particular theories see an ideology as a set of beliefs or prejudices that *mask* or *distort* reality, leading to the concept of false consciousness understood as deviance from a norm, requiring to be cured or corrected. The primary example is Marx's theory of the interest-based ideologies of those classes that have power, that is classes other than the proletariat, compared with the clear view that the socially powerless proletariat may have of history, society and even the natural world. Particular ideologies, then, are held to be not only distortions of reality, but *socially induced* distortions, arising from the class interest of the proponents and victims of the ideological beliefs. But why, asks Mannheim, exclude *any* social group from this analysis? The proletariat, Marx, the Communist Party, all have interests which are served by their own beliefs, and have interests in masking or deceiving themselves about the real situation; Marxism itself has a social function and is a social ideology. This brings Mannheim to his *total* theory of ideology, which turns 'unmasking' into a war of all against all: *all* beliefs about man and society are induced by social context, and have social functions. On an individual level, Freud's theory is an example of total ideology, in which the distinction between the normal and the deviant psyche becomes submerged in a total psychopathology; that is, in which the psychic characteristics of *all* individuals are held to be induced by their biological history.

Mannheim never succeeded in resolving the contradiction between the contrast of 'real' and 'distorted' belief on the one hand, and the notion that all beliefs distort the 'real' on the other. For if *all* beliefs distort, how can there be *true* beliefs about the real, and in particular how do we know there is a 'real' to be distinguished from the distortion? This reflexive argument certainly hits Mannheim's own theory, for this is quite clearly a social theory of the same kind as it refers to, and must therefore itself be socially induced according to its own principles. Mannheim's own suggestion for resolving this dilemma is that after all the intelligentsia is a disinterested class whose beliefs are minimally distorted. It is difficult to take this as other than a piece of frivolity, but if it is taken seriously it certainly contradicts the very notion of total ideology, according to which there cannot be beliefs about the real against which to measure degrees of distortion.

Mannheim attempts to save himself from the worst consequences of his theory by his second thesis, which explicitly excepts from the theory natural science, mathematics and logic. He presupposes without much argument that in these areas at least we can preserve the distinction between the body of true and grounded beliefs equated to knowledge on the one hand, and false belief or error on the other. This situation is of course not unsatisfactory for anyone who holds the positivist view that we can only have knowledge, and only make sense of the concept of truth, in the areas of science and logic (although this positivist belief itself, not being a scientific or logical belief, is hit by Mannheim's self-reflexivity, as Habermas, Gouldner and others have forcefully pointed out).[5] With the current decline of positivism, however, this second thesis of Mannheim's is now rejected by some of those who practice and reflect on the sociology of scientific knowledge, at the cost of reintroducing the reflexivity problems in more intractable forms.

In what follows, I shall denote as the *strong thesis* the view that true belief and rationality are just as much explananda of the sociology of knowledge as error and non-rationality, and hence that science and logic are to be included in the total programme.[6] It follows from the strong thesis that the sociology of knowledge is *symmetrical* in that it is not confined to the pathology of belief: to irrationality, or error, or deviance from rational norms. It rejects the view that *correct* use of reason, and true grounded belief, need no causal explanation, whereas error does need it.[7] There is, of course, a sense in which this is true but merely trite. For if an explanation of why someone got a piece of arithmetic right, for instance, is given as 'because he followed the rational rules', this can immediately be followed up by the causal question 'Why did he follow the rational rules?', to which the answer may be a combination of biological, psychological and sociological explanation. And this would be true even if it were also true that there are conclusive arguments for the necessity of the rational rules, for it is clear that few people (if any) in fact adopt such rules on pure grounds of rational necessity uncaused by any previous social history.

This much is generally conceded even by critics of the total sociology of knowledge thesis (it is conceded for example in Steven Lukes' review of Barnes' first book—a review that is otherwise critical of the thesis).[8] But the thesis is generally taken to imply something more, namely the view that even if there are rational rules that are ultimately independent of social (and perhaps biological) causation, at least no such rules can be appealed to as independent variables in a social explanation of knowledge.

To understand how such a radical thesis has recently been thought relevant to the historiography of science we must turn to recent developments in the philosophy of science. In particular the work of Quine, Kuhn and Feyerabend has led historians to supplement internal by social types of explanation of scientific ideas. This is not because any of these three writers have done much to explicitly encourage the study of sociology of science (in the cases of Quine and Kuhn if anything the reverse is true), but rather because of certain common features of their analysis which can be summarized by reference to the concepts of 'underdetermination' and 'incommensurability'. Quine points out[9] that scientific theories are never logically determined by data, and that there are consequently always in principle alternative theories that fit the data more or less adequately. He goes on to argue that any theory can be saved from falsification by apparently contradictory data, and conversely that any theory can be held to be falsified, provided sufficient adjustments are made in the extra-empirical criteria for what counts as a good theory. No theory can exactly capture the 'facts of the matter', even if it makes sense to speak at all of 'facts of the matter' outside the possibility of description by some theoretical conceptual framework or other. Moreover, since Quine holds that there is no separate category of *a priori* truth, it follows that these extra-empirical criteria are based neither on empirical nor rational foundations. Thus it is only a short step from this philosophy of science to the suggestion that adoption of such criteria, which can be seen to be different for different groups and at different periods, should be explicable by social rather than logical factors.

The suggestion is reinforced by the emphasis of Kuhn and Feyerabend on 'incommensurability'.[10] Conflicting scientific paradigms or fundamental theories differ not just in what they assert as postulates, but also in the conceptual meaning of the postulates and in their criteria of what counts as a good theory: criteria of simplicity and good approximation; of what it is to be an 'explanation' or a 'cause' or a 'good inference', and even what is the practical goal of scientific theorizing. All such differences are inexplicable by the logic of science, since they are precisely disputes about the content of that logic. The historian must make them intelligible by extra-scientific causation.

An example is to be found in the papers of Cantor and Shapin on the debate about phrenology in Edinburgh in the years 1803–28.[11] Cantor shows how the adoption of different theologies, natural philosophies, and even social philosophies, led different groups respectively to adopt or reject the theory of localization of brain function and its observable correlations in 'bumps' in the skull. Phrenology suggested a materialist unity of brain and mind, and was therefore opposed by mind-brain dualists. It also advertized itself as a supremely empiricist science, since its evidence was open to anyone to examine without long professional training in anatomy and physiology. Needless to say it was therefore opposed by the medical and scientific establishment. And these were differences that could not be settled by simple appeal to experience and experiment, since crucial experiments could be and were interpreted by each side in conformity with their own theory. With regard to this particular dispute the conflicting theories were unfalsifiable and incommensurable.

In his reply to Cantor's paper, Shapin is fully agreed about this logical situation. Where he differs is in the emphasis subsequently placed on the possibilities of social explanation. Shapin complains that Cantor has not gone on to explain the incommensurability and mutual misunderstandings in terms of their class origins and social aims, and he himself attempts to do so. In his second paper Cantor questions, not the propriety of such explanations in principle, but the particular ideological stance from which Shapin mounts his attempt—a stance that undervalues conscious intellectual choices in comparison with thoughtless commonsense and unconscious class motivations. Reduction of the cognitive to the social is, he thinks, in general hazardous for the historian, since at present cognitive influences on individuals are better understood than their underlying social causation. The dispute is not about the *presence* of external factors in scientific theorizing, but about their nature.

Historiographical points such as these will be taken up later in relation to the strong thesis. Meanwhile two other examples will be useful to show that it is not only disputes about fringe sciences like phrenology that lend themselves to extra-logical analysis. One such example is the study by Farley and Geison of the Pasteur–Pouchet debate about spontaneous generation.[12] This is also incidentally an excellent example of the way in which underdetermined theoretical commitments of such a general and flexible character as 'spontaneous generation' may at one time reflect the best interpretation of the data, at a later time not, and later still come again into favour, though with changes of definition and connotation. This is a feature of scientific theory that makes it hazardous to use current theoretical conceptions as legitimations for social or religious ideology, though it is clear from the history of science that they have often been so used.

Farley and Geison describe how, in the 1820s, Cuvier, the empirical scientist and political conservative, aligns himself against the postulate of spontaneous generation, which was associated with romantic *Naturphilosophie*, revolutionary politics, and philosophical materialism. By the 1860s, however, scientific legitimation had changed political sides and now favoured spontaneous generation. Now it is Darwinism that has scientific standing, and Darwinism requires spontaneous generation, certainly of the separate species during the course of evolution, and perhaps originally also of life itself. Darwinism is also perceived at this time to be inimical to religious order and political conservatism (in France, that meant maintenance of the Second Empire). It is Pouchet who champions Darwin and spontaneous generation, and finds himself obliged to argue their consistency with religion. Pasteur on the other hand is supported by official French science, which is anti-Darwin, and he is backed by Napoleon himself. Farley and Geison recognize that up to a point the arguments on both sides of the spontaneous generation debate are validly 'scientific', though Pasteur's use of experiments is sometimes questionable, and he seems to have privately modified his views towards acceptance of spontaneous generation as it would be required by Darwinism, without revealing this shift until the politically sensitive 1860s were past. The authors conclude that his behaviour is consistent with a greater degree of influence from external factors than in the case of Pouchet, but they do not deny that there were scientific 'facts of the matter' which came progressively to light. They also believe there are *historical* facts of the matter, for in a self-reflexive final paragraph they ask whether their own approach to the history has been influenced by their antipathy to Pasteur's social and political views, and claim that they have tried to set aside their own views and to seek objectivity critically!

There are two important features of this case. The first is the shifts in the *meaning* of spontaneous generation during the course of the debate, leading to incommensurability of the concepts used on both sides. Insofar as Pasteur asserted that life is not produced by 'ordinary chemical processes', he was right in the light of subsequent theories; insofar as he denied that it could ever be produced by sufficiently complex inorganic molecular processes he was wrong. The second important feature of the case is the use of scientific legitimations on opposite sides of the religious and political debates, depending on the historical circumstances of different periods. This suggests that sociology of knowledge will not issue in any simple correlation of types of methodology and types of social structure over history.

Another example of social analysis of an even harder science concerns the debate between Pearson and Yule regarding the best method of calculating regressions and correlations from statistical data. In 'Statistical theory and social interests: a case study',[13] MacKenzie shows how Pearson's use of the normal distribution gave him a model powerful enough for a predictive theory of heredity, and hence for applications in eugenics. Yule, on the other hand, rejected the normal distribution hypothesis, and stayed closer to the ordinary language classification of property-variables. His theory was therefore only weakly predictive, but it was sufficient for the applications he wished to make of it, which included problems about vaccination, but not applications to eugenics, which he rejected. Here there is incommensurability of *method* in developing theory, and of *cognitive interest*, that is of the aims of the theory in application. And the incommensurability occurs within a recent mathematical science in the ordinary course of theory development.

So far it may seem that nothing as strong as the strong thesis is required to justify historical investigations of the kind just exemplified. All that has been shown, it may be said, is that where logic and observation are insufficient to determine scientific conclusions, there historians may look to social explanation to fill the gaps. Recent analysis of underdetermination and incommensurability has shown simply that logic does not carry us as far in explaining the actual course of science in history as was once thought; but it does not show that social causation is symmetrically relevant to true and false logic and true and false science. Such a response demands more clarity about the claim to 'symmetrical' relevance, and hence about the meaning of the strong thesis. The first sense of 'symmetry' to be examined is the use that it seems proponents of the strong thesis have chiefly in mind: that is the view that there are no extra-natural and extra-social grounds of rationality and truth in the *a priori*, the analytic, or the necessary, and *a fortiori* that no transcendental argument or rational intuition can claim to have access to such grounds. It will be convenient to examine this sense of symmetry by means of its converse, that is by examining some types of rationalist argument which claim to find an irreducible and perennial rationality that is *not* explicable by natural or social causation.

Rationalist arguments against the strong thesis

The arguments I have in mind are those sometimes called 'transcendental', according to which there are certain necessary conditions upon a language or a belief system if there is to be any possibility of interpersonal communication within it.[14] The arguments can be distinguished into three types, according to whether they refer to necessary logical or scientific conditions, or are based on the common features of human life as a biological entity. Finally I shall consider the argument from the apparently self-refuting character of the strong thesis.

1. First let us consider the postulate that logical truth is an *a priori* necessity for language. It is sometimes argued that at least the concepts of negation, contradiction (the 'yes/no' distinction), and entailment must be present in order for us to know that another system of signs is a language at all. Therefore it is held to follow that all belief systems must have at least this much of a common logic. Various replies can be made to this point. First it is a positivist point in the sense that it presupposes that there can be no language unless we (now, or perhaps at some future time) understand it as a language. But suppose we come across Martians who are making visible or audible gestures and who are evidently in reasonable control of themselves and their environment, and yet we persistently fail to get started on translation of their sign system because no consistent sense can be made of any hypotheses we make about the signs for yes and no. Lest this seem implausible, we can think of it as the limit of cases where we *have* found sign systems

unintelligible (as in many anthropological and theological examples where the criteria of identity, for example, of men with birds, three persons in one, etc., do not answer to our criteria), but where we may eventually become convinced that these locutions are part of a language because we are able to *extend* our understanding and our language in unpredictable ways to give intelligibility. In such cases we have no evidence that there are *rules* for such extensions of understanding—they certainly do not always conform in obvious ways to the application of propositional logic. For example, some alien locutions come to be understood eventually as *metaphors*, comparable with our non-propositional use of language in poetic contexts. It is therefore never safe to claim that we have found necessary truths to demarcate language as such. To do so is to *define* the limits of language *a priori* in ways that can turn out arbitrary and unilluminating.

A second reply to the argument for logical necessities in language would be to point to various forms of non-standard logic that have been proposed in contexts other than that of discursive argumentation. For example, there is intuitionist logic in the foundations of mathematics, and proposals for three-valued logic in quantum physics.[15] This is too large a question to go into here, except to say that the possibility of different basic logics is not by itself very cogent, because it may be said that the examples we know of are all parasitic on standard logic for their learning and defence. It is perhaps safest to take from this argument only the Quinean point that some but not all logical truths must be retained from system to system, but that we cannot tell *a priori* which may turn out most convenient to retain in any given circumstances.

A more important, and it seems to me decisive, argument against the rationalist's position on the logic of language is to point out that all it could possibly prove would be a purely formal similarity of logical structure between belief systems. If language is to convey information, then it does necessarily follow that it contains at least some binary distinction corresponding to yes/no, agreement/disagreement, true/false, that is, it contains elementary 'bits' of information. But this says nothing whatever about the *content* of formal logical principles, that is, the way they classify the world of discourse in any particular language. For example, the so-called 'laws of thought' do not determine any particular content. The various candidates for necessarily false statements that have been canvassed (e.g. 'a red object is not coloured') are certainly not formal contradictions in the sense required if the law of contradiction is to be necessary in every natural language. It is perfectly possible, and indeed it occurs, that particular alleged synthetic falsehoods in our language are not such in other languages, and that we can with ingenuity and hard work both understand and eventually translate them satisfactorily into our language. Similar arguments could be developed for the concept of entailment and other alleged necessary logical truths. Whenever they are necessary they will be found to be empty, whenever they are informative about the content of another language they will be found to be contingent.

There are other arguments for the necessity of some logical truths which depend on a greatly extended use of the notion of 'logic' as a kind of 'natural rationality' that must be exhibited in all languages to which we 'catch on' rather than merely become conditioned to by habit. The character of such natural rationality remains however extremely obscure. Is it a form of Kantian transcendent, such as is suggested by Strawson's descriptive metaphysics of space and of persons? Is it like the interlingual structure of Chomskyan linguistics? Or is it more like the habit of making simple inductive generalizations, which surely *is* the result of psychological and physiological conditioning in the sort of contingent world we find ourselves in? None of these alleged 'content' logics or rationalities have been adequately developed as candidates for the necessary rationality of a language, and in any case it would appear from our conclusions thus far that though they may be universal in scope over all existing languages, they are contingently and not necessarily true.[16]

2. This brings us to the second major claim of rationalist argument, namely that there are certain prerequisites of elementary empirical science that must be necessary conditions of any belief system. There must, it is argued, be some empirical reference in common of some general terms in all

languages if communication is to occur. On this point I will only refer to Quine's powerful arguments to the effect that no conclusions about what the referents are, or how they are classified, can be drawn from success in interpersonal communication based purely on linguistic behaviour. It is further argued that some knowledge of the environment of an elementary scientific kind is necessary in all societies for human survival. This is true, but it does not at all follow that this common 'knowledge' has to be expressed in language or even form part of an explicit system of beliefs. Here even Quine's thought experiments require too much—for instance learning that 'gavagai' has to do with rabbits in the alien society presupposes that the aliens have to *talk* about such directly observable things as the quarry of their hunts, contents of their cooking pots, etc. But this does not seem to be a necessary consequence of having a language. Suppose a certain society talked only about more important things like their relation to the spirit world, and not at all about the mundane processes of keeping alive, procreation, and so on. We could not conclude that they had no language, though for our secular society it might be very difficult for us ever to come to understand that language, because it might have no referents in common with ours.

A further rationalist argument from the alleged commonality of science is to the effect that science provides accumulating predictive control of the environment, and that its evidently privileged position in this respect over all other known belief systems is a universal touchstone of rationality. But it should be remembered that many arguments of this kind still take a pre-Kuhnian view of the relation between scientific *theory* and pragmatic predictive success. Even if the notion of 'success' itself is agreed upon, it does not follow that either the theoretical aspects of science, or their observation languages (which are themselves theory-laden), share the progressive and accumulative characteristics of scientific practice. There are no grounds for supposing that two societies with comparable practical science necessarily share any particular presuppositions of empirical reference or theoretical ontology. Many such candidates beloved of rationalist philosophers in the past, such as alleged *a priori* properties of space, time, matter and causation, have been shown to be modifiable in modern physics without loss of empirical content. Again the Neurath–Quine metaphor of replacing the planks in a floating boat is relevant; no doubt we shall be able to see *post hoc* that there is some overlap in the presuppositions of successive theories and in the extensions of some of their respective general terms, but none can antecedently or postcedently be identified as necessary. There is no material content of language or theory that is *dictated* by the empirical.

3. Finally, there is a group of arguments from the common biological basis of human life to the necessity of some 'rationality' in common between all belief systems. Among such arguments are Dilthey's attempt to articulate a universal *Lebensphilosophie*, Malinowski's functional analyses in terms of human biological needs, and even Winch's reference to the universal features of birth–life–death which serve as the motive for some ritual observance in all societies. For those with more empiricist preferences, we might also point to Quine's interlingual observation or 'occasion' sentences, and the 'stimulus meanings' which provide a universal type of causal link between observation sentences and the environment. In regard to these types of argument, Barry Barnes concludes that where epistemologists seem to be developing *a priori* theories of induction and other forms of rational inference, what they are in fact doing is describing processes of natural inference and learning—processes that are of essentially the same kind as the conditioned reflexes and more complex learning devices which men share with animals, albeit in more advanced evolutionary forms.[17] Even if all these features of human rationality are accepted as universal, as in some sense they surely must be, they do not give grounds for concluding that such universal features are necessary truths. At most it might be suggested that the universality and *biological* necessity of 'natural inference' gives grounds for pragmatic concepts of 'truth' and 'knowledge' which will have to form part of the claims to knowledge in any culture.

4. The most powerful rationalist argument against the possibility of a strong sociology of knowledge thesis is undoubtedly the so-called argument from self-refutation. The usual form of this argument goes as follows: Let P be the proposition 'All criteria of truth are relative to a local culture; hence nothing can be known to be true except in senses of "knowledge" and "truth" that are also relative to that culture'. Now if P is asserted as true, it must itself be true only in the sense of 'true' relative to a local culture (in this case ours). Hence there are no grounds for asserting P (or, incidentally, for asserting its contrary).

This easy self-refutation is fallacious, for it depends on an equivocation in the cognitive terminology 'knowledge', 'truth', and 'grounds'. If a redefinition of cognitive terminology as relative to a local culture is presupposed in asserting P, then P must also be judged according to this redefinition. That is to say, it is fallicious to ask for 'grounds' for P in some absolute sense: if P is asserted, it is asserted relative to the truth criteria of a local culture, and if that culture is one in which the strong thesis is accepted, then P is true relative to that culture. We cannot consistently ask for absolute grounds for accepting either P or the strong thesis.

Now of course this is not a conclusive argument for accepting the strong thesis. There is no such conclusive argument. It does not *follow* from ideological analysis of other people's beliefs and of certain aspects of our own that the rest of our own system is culturally conditioned and culturally relative. What the argument from sociology has done is to suggest that we *shift* our concept of 'knowledge' so that the alleged refutation becomes an equivocation. This shift is the essence of the strong thesis: knowledge is now taken to be what is accepted as such in our culture.[18] This may appear to be a circular support of the thesis by means of redefinition of its terms. But suppose we treat the thesis, not as a demonstrable conclusion from acceptable premisses, but rather as a hypothesis in the light of which we decide to view knowledge, and consider whether its consequences are consistent with the rest of what we wish to affirm about knowledge, and whether it does in the end provide a more adequate and plausible account than the various rationalist positions we have found questionable. Consider the following alternative situations:

(i) We have some absolute criteria of knowledge in terms of which we can make absolute evaluations of belief systems including other parts of our own, and

(ii) We have culturally relative criteria of knowledge in terms of which we can make relative evaluations of belief systems including other parts of our own.

Option (ii) is undoubtedly self-referential, but it is not self-refuting. I think Steven Lukes, for example, tacitly recognizes this in his arguments for independent criteria of truth in his 'On the social determination of truth'. He says:

> Any really hard-boiled relativist would just reject these arguments [for independent truth] as themselves relative, but to do so he must realize the full implications of the pluralistic social solipsism his position entails. . . . Are the truths of [a group's] beliefs and the validity of their reasoning simply up to them, a function of the norms to which they conform? I maintain that the answer to this question is no—or at least that we could never know if it were yes; indeed, that we could not even conceive what it would *be* for it to be yes.[19]

There does not seem here to be the suggestion that the answer yes is self-refuting, just that, firstly, it would be excessively unpleasant, and secondly, that we could perhaps not even conceive what an affirmative answer could mean. With the first point I will not quarrel, except to say that to characterize it as making truth and validity merely 'up to us' is a misdescription, for if something is a function of a cultural situation it is certainly not wholly under our individual control, and therefore not just 'up to us'. And Lukes' second point about inconceivability seems to be just wrong: his arguments for it depend only on the unobjectionable view that we could not understand the other group's language or even know that they are asserting or arguing at all, unless they have some criteria of truth and validity in common with us. But even if this were true, it does not show that these criteria are in any sense external or 'absolute', only that they are relative to at least our pair of cultures, rather than to just our culture.

In my discussion of types of rationalist argument for

necessary truth in the previous sections, I have indicated my belief that no such arguments are cogent. If this is correct, it follows that there is no argument capable of establishing (i) rather than (ii). Our own criteria of knowledge include as many of the norms of logic and science as we normally adopt in our culture. The strong thesis entails no sort of restraint upon their use (I am, for example, attempting to use them correctly now). The strong thesis does indeed suggest that, since they are culturally, and perhaps also biologically, conditioned, we are in fact less free to change them as a matter of individual decision than some relativists might have thought. But if there are problems about which criteria are more central than others, and whether to add to or subtract from some of them, every society possesses various ways of resolving such disputes. In the absence of necessities the claim that there is a distinction between (i) and (ii) is a metaphysical distinction in the sense that we cannot derive any contradiction with our reasoning behaviour from either standpoint.

We might ask, for example, whether recognition of the self-reflexive character of the sociology of knowledge occasions any embarrassment or inconsistency in the work of historians and sociologists of science. Farley and Geison, as I have remarked, raise the question explicitly, and rest their reply on the assumption that there are, in our society, recognized criteria for 'objective history', which they hope to have satisfied. In other cases the question arises in debate between different historiographical principles. For example, we have seen that it is the 'rules of the game' in writing history of science that are principally at issue between Shapin and Cantor, one adopting the standpoint of Marxist historiography ('change is to be traced to the class struggle'), the other adopting a more traditional, descriptive, type of historiography. But of course, insofar as both have put forward their work in a particular professional community, namely their colleagues and the journals of English-speaking historiography, they have necessarily accepted the common constraints of that community—constraints of evidence and argument that are more general and more comprehensive than the constraints of their own particular standpoints. Why *those* common constraints are adopted in the twentieth-century Western world is a further question for social explanation. But no question of a logical regress arises, because no explanations are ultimate. Somewhere explanation stops at the point where, temporarily perhaps, it is not questioned by the relevant local consensus. What my arguments against the rationalists have been designed to show is that that stopping point is not in the realm of *a priori* and necessary truth.

The strong thesis and epistemology

The previous section has been relevant to an interpretation of the strong thesis that would exclude necessary truth as an ingredient of explanation in the sociology of knowledge. If any of the contrary rationalist arguments are cogent, then the strong thesis would undoubtedly fall. On the other hand, if as I have suggested they are not cogent, the strong thesis survives, but we have to consider whether we have learned much more about its nature and its consequences. In the first place it does not follow that there *is* no necessary truth, only that if there is it cannot provide analytic premisses for explanations in the history of science. Of course the strong thesis is not alone in suggesting that we can never validly claim to *know* in the sense of knowledge that implies that what is known is true. Any view which rejects the possibility of secure foundations for knowledge faces the consequence that such applications of cognitive terminology can only be made tentatively and relatively to whatever insecure foundations it seems plausible to adopt.

Neither does it follow from such views or from the strong thesis that all talk of truth, reasons and knowledge becomes inapplicable and empty. We can certainly still give cognitive terminology a use, although it will be a use different from that current among rationalist philosophers. According to the strong thesis, what epistemologists study is the rules accepted as rational in their own society, much as mathematicians study the relations between and consequences of postulates adopted arbitrarily with respect to such external criteria as the empirical, the self-evident, or the logically necessary. Every society makes a distinction between some set of cognitively binding rules on the one hand, and mere social conventions on the other. Rules governing activities like counting and discursive argument are different from conventions like driving on the left, and even from symbolic expressions like crossing oneself from left to right rather than from right to left. And within the set of 'rational' rules, every society distinguishes between norms and deviations, correctness and error, truth and falsehood. Hence every society can have its epistemologists and its standard ways of using cognitive terminology. The function of epistemologists to make these distinctions explicit and to study their interrelations is both important and not directly sociological. And it is not obviously undermined by further sociological investigation of the causation of these rules and comparisons between them and the rules of other societies, nor by the fact that such investigation uses some of the very same rules that the epistemologist studies. We carry out such circularly supported investigations all the time.

In such a new construal of cognitive terminology, rules of argument and criteria of truth are internal to a social system, or perhaps to a set of social systems, but this account does not remove the motivation for epistemological studies nor emasculate philosophical theories. Such consequences would only follow for those who retain a rationalist theory of knowledge, not for those who accept the redefinition of 'truth' and 'rationality' implied by their status as internal to given societies.

Reference to some case histories will help to clarify the place of internally defined 'rational rules' in the historiography of science. A particularly important point at issue is the question of the demarcation of 'science' itself. If it is accepted that science is not distinct from other kinds of belief system in respect to the *a priori* rationality of its basis, does it have any other distinguishing features that the historian of science should recognize, even if only to demarcate his own subject matter?

In his defence of the strong thesis, Barry Barnes has argued that all attempts to find demarcating criteria, that is, necessary and sufficient conditions for a belief system to be a science, have failed.[20] These failures include all verifiability and falsifiability criteria, and all specific appeals to experimentation and/or to particular kinds of inductive or theoretical inference. At best, he argues, the concept 'science' must be regarded as a loose association of family resemblance characteristics involving, among other things, aversion to all forms of anthropomorphism and teleology, and consequent tendencies to secularism, impersonality, abstraction and quantification. Moreover, we must not impose our own scientific criteria on the past; the subject matter of the historian of science can only be demarcated by recognizing what it is in the past that exhibits causal continuity with present science.

While accepting Barnes' general framework, however, it is not inconsistent to point out that historians of science need to recognize and to explicate various sorts of rules of scientific inference as these were consciously or unconsciously adopted both in past and present. For example, in his article 'Statistical theory and social interests' mentioned above, MacKenzie accepts the demarcating criteria of 'technical interest' for science, that is the requirement of predictability and potentiality for control of nature, but points out that within this general interest, different specific goals for science may manifest themselves in different types of application. And he recognizes that individual 'convictions' (which may presumably be swayed by rational argument) can be distinguished from social constraints, as when, for example, he points out that Yule and his followers did not form a tightly knit professional school or exhibit a common social background and orientation. None of this is inconsistent with the strong thesis, for the 'technical interests' and the 'rational convictions' can all be understood to be rules adopted within the society under investigation.

Again, in his debate with Cantor over phrenology, Shapin distinguishes some non-technical interests of the phrenologists and their opponents, for example social reformism and appeal to democratic common sense. But neither he nor Cantor have any difficulty in distinguishing these from what Cantor calls 'cognitive' factors. The issue between them is not that of the possibility of demarcating different kinds of interest and correspondingly different kinds of science or pseudo-science, but rather what the causal and reductive relations between these different interests are, and what ought to be taken as 'explanation' by the historian.

What all this shows is that, while adopting the strong thesis with regard to science, historians both can and should make distinctions between the kinds of socially evolved rational rules within which they and their actors are working. There are no uniquely satisfactory demarcating criteria for science, but there are various sets of criteria that can be found to be operating in different circumstances. If we wish to understand why science appears to us to be necessarily progressive, and yet wish to take account of the radical changes through which science passes and in which there seems to be no theoretical progress, it may be that the technical model of increasingly successful predictive control is the best one. If on the other hand we are interested in the congruencies and continuities of natural and social science we may need to highlight other, more theoretical and hermeneutic features which provide a socially legitimating function. Not only are there several viable models for our science, but also every culture's image of its own science is different. The historian of science has to take account both of his actor's own self-image, and of that in his own society, as well as of those in between which span the historically continuous process that counts as the history of science. So the strong thesis does not imply any sort of homogeneity among various sets of conventional rules, rational and other, just because it recognizes them all as socially conditioned.

The strong thesis and causality

Further clarification of the strong thesis is required in relation to use of the terms 'explanation', 'causation', 'determination' and 'conditioning'. What kinds of social causation of knowledge does it imply?

First it must be emphasized as a general principle of the philosophy of science that causal explanation and the existence of general laws does not entail causal determinism. Determinism is the characteristic of a theory according to which from a complete description of the present and perhaps some past states of a system, all future states can be precisely and uniquely calculated. No natural science has ever validly claimed the world to be deterministic in this sense, not even Newtonian mechanics, if only because the complete information required to test such determinism can never be available. In any case, modern physics describes the world as in principle indeterministic, and gives explanations and predictions only in terms of statistical laws. It is none the less a causal science for that. Moreover, there are other natural sciences such as cosmology, geology, and evolutionary biology that contain even more loosely specified causal correlations and trends. In some cases significant individual events are unpredictable, and yet cause radical changes in the future course of the system: for example, individual particle disintegrations or individual mutations. All this does not prevent us speaking of causal laws and explanation in natural science.

At least the same variety of non-determinist causality must be expected in social theories. In considering the possibility of the sociology of knowledge, the distinction between causality and determinism is particularly important. The social explanation of knowledge may be causal but not deterministic, that is, it may restrict the possibilities without determining any single one of them. It might be tempting to argue that the gaps in causal explanation could then be filled by further explanation in terms of rationality: 'He was trained by such and such learning procedures to do arithmetic, but what completes the explanation that he got his sums right was the rational correctness of his procedures.' But such a suggestion would be inconsistent with what I have taken the strong thesis to assert; moreover I have indicated in section 2 why I do not believe that such independent rational factors have any part to play in explanation.

A much more important point concerns the consequences for the sociology of knowledge if social theory had to presuppose that the world is deterministic. A deterministic world is a world without free choice. I take a 'free choice' to be such that there are two or more possible and mutually exclusive outcomes of the event of choice in the physical or social world or both, all these outcomes being consistent with all physical and social theories, however complete and adequate these theories may be. After the event of free choice the world is in one and only one of the possible outcome states. I do not wish to analyse the concept of free choice further, except to say that the foregoing definition seems to me to be the minimum that is required for the ordinary notion of free choice, and that if there are no occurrences of free choice in this sense, particular difficulties are caused for the sociology of knowledge (as also, I think, for ethics, but that is a further large question). For suppose sociology of knowledge were a total and deterministic theory. Then the fact that in some societies (ours among them) some people adopt forms of the sociology of knowledge thesis, would be a determined fact. This would make it pointless to speak of this approach to knowledge as being freely chosen after consideration of reasonable, but not logically conclusive, inferences from philosophical and empirical premisses. But most sociologists of knowledge do put forward their theses as positions to be argued for, and as persuasive but not coercive of beliefs. The consequence appears to be a *reductio ad absurdum* of the sociology of knowledge, and indeed of all argument whether understood in the rationalist's or the sociologist's sense. Rather than trying to provide stronger arguments for its unpalatable character, I prefer to take the realistic course of dropping the assumption of determinism. If social causality is understood in the weaker senses that are common in natural science, this kind of self-reflexive refutation of the strong thesis does not arise.

There are two further points to be considered, however. The first is the question whether, consistently with the strong thesis, it is possible to engage in the kind of argument I have just outlined at all. What becomes of 'free choice after consideration of reasonable but not logically conclusive inferences' for example? Glossed into the forms of cognitive terminology required by the strong thesis, this becomes 'free choice constrained but not determined by local rational rules'. So far as causal explanation is concerned, such free choice will appear as an individual chance event, just as it does for a non-determinist with regard to explanations within, for example, brain physiology or social psychology. I have said that I do not propose to go further here in trying to elucidate the notion of free choice, but it is relevant in the present context to remark that I would see any further elucidation in terms of moral events rather than rational rules. That is to say, I believe sociologists of knowledge are right in denying the possibility of purely rational premisses as independent variables in explanation, but wrong if they deny the possibility of uncaused moral choice.

The second point about determinism concerns an argument that a persistent determinist might use against the strong thesis. If the strong thesis is adopted along with determinism, he might say, we appear to arrive at a *reductio ad absurdum* according to the argument developed above. So why not drop the strong thesis rather than determinism? This suggestion does not work, however, just because the *reductio* argument applies to all forms of rational inference, and not just to the strong thesis. In a strictly determinist world all rational inference is vacuous unless, like Descartes' pineal gland, it has some effects in the physical world, even if these are only the products of speaking and writing the expressions of the rational argument. In a determinist world all these physical effects are themselves determined. Thus dropping the strong thesis does not by itself help the determinist; it merely adds the claim that the determinants of rationality are social rather than merely physical. It is odd that rationalist philosophers often view with equanimity the possibility of a strictly determinist *natural* world, while reacting with extreme distaste to the suggestion that there are explanatory laws of the sociology of knowledge. Would they rather the determinants of thought be purely physical than mediated through social relations?

Substructural explanation and historical law

Short of determinism, there are many types of causal explanation that might be appealed to by the sociologist of knowledge. A particularly important question concerns the possibility of the causal reduction of superstructural systems of ideas and cultures, including knowledge, to the substructure of socio-economic systems. Marx's original motivation towards the sociology of knowledge sprang from his belief that the socio-economic structure does form such a reductive basis for explanation of ideology and culture, and it is clear from many of the studies in social history of science to which I have referred, that they share this presupposition, and would take the strong thesis to include it.

We may, however, follow Durkheim and Weber in taking a more liberal position on this question, and detaching the strong thesis from substructural reductionism. The strong thesis as I have explicated it requires only that all aspects of social structure, including its cultural manifestations in ideas, beliefs, religions, art forms and knowledge, constitute interlocking systems of causality. Sometimes class structure may be a causal factor in ideology; sometimes ideological exhortation may be a cause of social change. Every historical case has to be examined on its individual merits. Sometimes 'cognitive factors' such as local rational rules may act as independent social variables, as Cantor and even MacKenzie and Farley and Geison suggest. Sometimes the domain of independent rational choice is small compared with an individual's social goals or causal factors due to class background, contemporary social structure and social conflict. This is the view of sociology of knowledge classically exemplified in Weber's *Protestant Ethic and the Spirit of Capitalism*, and in Durkheim's recognition of symbolic and ideological systems as *sui generis*.[21] It is, of course, not surprising that those who adopt a Marxist orientation towards substructural reduction, should also tend to be social determinists, since they are not troubled by the kinds of problems about individual choice that I discussed in the previous section. There are, however, other writers who do not explicitly link the strong thesis with Marxist reductionism, but who nevertheless appear to drift unthinkingly and unnecessarily towards both substructural reductionism and strict social determinism.

A further question about the character of social causation may be expressed in terms of what I shall call the 'intensive' and the 'extensive' approaches to history of science. Historians in the extensive mode tend to make general hypotheses and to look for general correlations across societies, across periods and across sciences. Such correlations have been suggested between, for instance, social class of origin, and propensities to positivism or realism, or atomism or field theory, catastrophism or uniformitarianism, individualism or holism. Hypotheses with more general social application are also now being tested in the history and philosophy of science.[22] Certainly such attempts have not yet been sufficiently developed for it to be possible to judge their success, but first impressions do not seem favourable. For it is evident that at different periods of history and in different circumstances, line-ups of allegiance between social and ideological factors are liable to change—we saw one such example in the debate on spontaneous generation. Another comes from the implications of physical mechanism in the seventeenth and eighteenth centuries. In the eighteenth century, for example, Priestley is the religiously dissenting marginal man who favours dynamic theories of matter against the pure mechanical law favoured by the religious establishment. In the seventeenth century, on the other hand, it was sometimes the new radical secularists who were associated with mechanical atomism against the scholastic establishment with their multiplicity of essential forms and qualities of matter.[23]

Such examples suggest that at least there are no simple cross-period correlations to be found on the conceptual surface of history. Whether deeper structures like those suggested by Foucault are to be found requires further historical investigation. But whatever the outcome, intensive studies of the history of particular periods with their multifaceted causal linkages will still be an essential prerequisite of the history of science. This approach has been well exemplified in all the cases we have so far considered. Another example that illustrates its implications is the study by MacKenzie and Barnes of the Pearson–Bateson controversy between the biometric and Mendelian theories of heredity.[24] Here the authors argue for a general correlation of the professional middle-class Fabian gradualism of Pearson's social group, with their interest in eugenic 'engineering' and their consequent positivist and precautionary aversion to speculative and ill-grounded theory. On the other hand, Bateson's social background was the Cambridge élite, with an almost romantic conservative affinity for non-mechanical, holistic theories, aversion to social engineering, and rejection not only of social Darwinism but also of the natural gradualism apparently implied by Darwin's theory. The historical method here is not general induction, for no claim is made that these social characteristics and these scientific propensities will be found correlated at all times and places. There are indeed in the MacKenzie–Barnes paper some attempts at Baconian-type arguments about presence and absence of factors in different particular cases: for example, scientific socialization and training are rejected as sufficient causes of differences of subsequent scientific style, on the grounds that some individuals cross over from one group to the other. On the other hand, single individuals are not allowed to be counter-examples to the main thesis, as when the authors remark that Pearson's collaborator Weldon did not reveal any commitments in politics, and that in general *there are no necessities* in an individual's chain of affiliations. The analysis involves rather 'the ongoing practice, ideology and institutional structure' of 'coherent social groups',[25] where the social groups are evidently ideal types in Weber's sense rather than actual entities with intrinsic causal power.

All this suggests that the causality sought in such cases has more to do with the historian's perceptions of ideological relevance and irrelevance than with mere external correlations of factors in particular cases. For example, MacKenzie and Barnes judge that features of scientific training are possibly relevant to subsequent scientific style, and that Fabian affiliations are possibly relevant to social and natural Darwinism. In a case as close to us historically and culturally as this one, where we share conceptions of rationality, the historian can be guided by his own 'understanding' of the relevance judgments and inferences of his actors. In cases more remote in time and space, this kind of internal understanding becomes more problematic.[26]

The intensive approach is not, however, the same as a non-causal method of *Verstehen*. As Weber himself clearly saw, causal correlations may be arrived at by comparisons of agreements and differences between particular analogous historical cases, even where generally applicable causal laws cannot be found.[27] This is, moreover, in line with Mill's own use of his methods of agreement and difference, which were not intended to yield universal inductions, but rather 'eductions' from particular to particular among a set of cases which are sufficiently analogous to each other. In the kinds of cases we have been considering, the evidence might, for example, be sufficient to permit causal generalizations from individual to analogous individual in similar historical circumstances, but not from one historical situation to another.

A final case may serve to illustrate this point. In his paper entitled 'From Galvanism to electrodynamics: the transformation of German physics and its social context',[28] K. L. Caneva has contrasted the respective styles of two generations of German physicists in the 1820s. The older generation maintains the traditional, concrete, qualitative style of Galvanism; the younger generation works in an increasingly professionalized science, and espouses the abstract, quantitative and rationalizing style of the new electrodynamics. Caneva even has a general psychological hypothesis: he assumes 'that a mechanism exists by which a perception of social relations can be transformed into a metaphysical conception of the nature of knowledge in general'. But it would be hard to find evidence (nor does he claim) that such a mechanism, once found, would be seen to operate in the same way in different historical situations, yielding law-like correlations across time. A correlation is identified in this case between conservative tradition and the empirical approach to science on the one hand, versus reformism and programmes for the planned society, together with abstraction and the hypothetico-deductive method which *imposes* form on nature on the other. This is a similar correlation to that found by Mackenzie with regard to Yule's empiricism versus Pearson's statistical abstraction and reforming eugenics. It is, however, the *converse* correlation that is found by MacKenzie and Barnes in their comparison of Bateson and Pearson, for there Pearson is still the reforming eugenicist, but the need for caution in application is said to yield *aversion* to the sort of speculative theory that is characteristic of Bateson's Mendelianism. The Caneva-type correlation is also found by Farley and Geison with regard to Cuvier's empirical conservatism, but is reversed with regard to Pasteur versus Pouchet. The meaning of terms like 'empiricism', 'abstraction', 'conservatism', 'reformism' are too dependent on historical context to yield general laws: when they are made precise, diachronic correlations turn out false; when they are left vague, correlations become vacuous.

Conclusion

I have suggested that the strong thesis of sociology of science is not as strong as some of its proponents have implied and some of its critics have objected. I have taken the thesis in what I understand to be the spirit of Mannheim's 'total theory of ideology', and Bloor's 'strong programme', namely that rational norms and true beliefs in natural science are just as much explananda of the sociology of science as are non-rationality and error. In the last three sections I have argued that this thesis, along with all other epistemologies that reject the possibility of absolute rational grounds for knowledge, implies that cognitive terminology cannot be applied in an absolute sense. But the thesis does not imply that cognitive terminology loses its use, merely that it has to be explicitly redefined to refer to knowledge and truth claims that are relative to some set or sets of cultural norms. These might even be as wide as biological humankind, but if so, they would still not be rendered absolute or transcendentally necessary in themselves. The strong thesis does not imply, however, that there is no distinction between the various kinds of rational rules adopted in a society on the one hand, and their conventions on the other. There may be hierarchies of rules and conventions, in which some conventions may be justified by argument in terms of some rational rules, and some subsets of those rules in terms of others. None of these possibilities imply that rational rules go beyond social and biological norms to some realm of transcendent rationality. Finally, with regard to causality, the strong thesis does not imply social determinism, nor any particular privileged direction of causality from substructure to superstructure, or indeed the converse. Nor is it implied that there are any universal laws of scientific and social development applicable generally across time. All that is implied is the possibility of finding some correlations, amounting to historical explanations in particular cases, between types of scientific theory and particular social provenance.

It may be felt that the 'strong' thesis has now become so weak as to be indistinguishable from something any rationalist or realist could accept in regard to the development of science. It may be objected that the kinds of rationalist arguments I referred to in section 2 are not intended to establish any extra-empirical or extra-social grounds of rationality, but only to elucidate the logically necessary presuppositions of beliefs that we (in our culture) actually accept about the nature of language, communication, and science—beliefs to which we exhibit our commitment in interpersonal relations. I have already left open the possibility that there may be such 'transcendental' analysis of our presuppositions; only objecting that any such conclusions cannot be intrinsically necessary, but are dependent on contingently held beliefs and beliefs about contingencies. The 'rationality' thereby argued for is no stronger than these contingent beliefs.

It is here perhaps that the real issue dividing rationalists and sociologists of knowledge begins to emerge. Associated with the moderate form of rationalism just described is often a belief that analysis of 'our' language, rationality and science will reveal the presuppositions of any possible language, rationality and science. In other words, rationalists are not impressed by the suggestions coming from the history of science and philosophy, the anthropology of other cultures, and Marxist analyses of ideology, that 'our' language may be relatively culture-bound. Those who adopt the strong thesis, on the other hand, even in the severely modified version I have presented here, are those who accept at least some of the relativism implied by those critical and comparative disciplines. Seen like this, the debate is not so much between rationalists and relativists, as between evolutionists and critical or hermeneutic theorists. For those who ground their faith in universal rationality on a contingent belief that our language and science are somehow the high points of the historical evolution of ideas, are in effect progressive evolutionists with regard to ideas, while those who believe that social and historical analysis can provide a valid critique even of our own presuppositions, are nearer to the tradition of hermeneutics. And hermeneutics depends neither on uncritical analysis of our language as if it were language as such, nor on the incommensurable relativity of languages and forms of life, but on the assumption that cross-cultural understanding and self-reflexive critique are both possible and illuminating.

Notes

1 A first draft of this paper was read at the conference of the British Society of the History of Science at Southampton, in July 1976. I am grateful for the comments then made, and for several subsequent discussions with Barry Barnes and David Bloor. I also owe to David Bloor several of the references to literature in sociology of science that are discussed below.

2 For references see the essays in M. Teich and R. M. Young, *Changing Perspectives in the History of Science*, London, 1973; also Charles Webster's *The Great Instauration*, London, 1975.

3 P. Forman, 'Weimar culture, causality and quantum theory, 1918–1927; adaptation by German physicists and mathematicians to a hostile intellectual environment', *Historical Studies in the Physical Sciences*, ed. R. McCormmack, vol. iii, 1971, 1.

4 Cf. also R. G. A. Dolby, 'Sociology of knowledge in natural science', *Science Studies*, vol. i, 1971, 3.

5 J. Habermas, *Knowledge and Human Interests*, London, 1972, ch. 1; A. Gouldner, *The Coming Crisis of Western Sociology*, London, 1970. Parts I and IV.

6 In thus characterizing the strong thesis I am following Bloor's four tenets for his 'strong programme in the sociology of knowledge' in *Knowledge and Social Imagery*, pp. 4–5. While I agree with many of his subsequent arguments, I give below a somewhat different interpretation to his concepts of 'impartiality' and 'symmetry'.

7 The converse has often been presupposed in history of science. For example the title of Bachelard's book *The Psychoanalysis of Fire* (English edn, London, 1964) exactly describes its contents as concerned with the 'irrational' accompaniments of the 'rational' developments of a theory of heat in the eighteenth century, studied as a mere appendage to scientific history.

8 S. Lukes, *Social Studies of Science*, vol. v, 1975, 501.

9 'Two dogmas of empiricism' in *From a Logical Point of View*, Cambridge, Mass., 1953, p. 20, and *Word and Object*, New York, 1960, chs 1 and 2. Subsequently Quine has moved to a somewhat less radical position in which currently accepted scientific knowledge is given privilege as the best we can know: cf. *Ontological Relativity*, New York, 1969, particularly chs 2, 3 and 5.

10 Principally T. S. Kuhn, *The Structure of Scientific Revolutions*, 2nd edn,

Chicago, 1970; and P. K. Feyerabend, 'Problems of empiricism', *University of Pittsburgh Series in Philosophy of Science*, ed. R. G. Colodny, Englewood Cliffs, N.J., vol. ii, 1965, and vol. iv, 1970.

11 G. N. Cantor, 'The Edinburgh phrenology debate: 1803–1828', *Annals of Science*, vol. xxxii, 1975, 195; S. Shapin, 'Phrenological knowledge as the social structure of early nineteenth-century Edinburgh', *ibid.*, 219; and G. N. Cantor, 'A critique of Shapin's social interpretation of the Edinburgh phrenology debate', *ibid.*, 245.

12 J. Farley and G. Geison, 'Science, politics and spontaneous generation in nineteenth-century France: the Pasteur–Pouchet debate', *Bulletin of the History of Medicine*, vol. xlviii, 1974, 161.

13 D. MacKenzie, 'Statistical theory and social interests: a case study', *Social Studies of Science*, vol. viii, 1978, 35.

14 I am not of course in this context trying to give anything like an adequate account of arguments of this type, which are the subject of an ever-expanding literature in current analytic philosophy. I intend only to point to various types of argument that have been exploited, particularly in a context not dissimilar to the present one, namely the anthropological problem of explicating the rationality of alien thought systems in relation to our own. See particularly papers by P. Winch, E. Gellner, A. MacIntyre, S. Lukes, M. Hollis and B. Barnes in *Rationality*, ed. B. Wilson, Oxford, 1970; *Modes of Thought*, ed. R. Horton and R. Finnegan, London, 1973; and M. Hollis, *Models of Man*, Cambridge, 1977.

15 Cf. S. Haack, *Deviant Logic*, Cambridge, 1974.

16 Robin Horton has suggested a Strawsonian theory of some such bio-social universals across all natural languages, but this is a far cry from Kantian transcendentalism, and is surely not a hypothesis that need be rejected out of hand by any sociologist of knowledge unless he adheres to a Marxist prejudice in favour of environmental as against hereditary causation ('Material-object language and theoretical language: Toward a Strawsonian sociology of thought', privately circulated).

17 *Scientific Knowledge and Sociological Theory*, p. 25; *A Theory of Classification*, forthcoming.

18 In his reply to the objection from self-refutation (*op. cit.*, p. 13), Bloor does not quite come to grips with this point. He argues that all such objections depend on the hidden premiss that what is socially caused is false; the strong thesis is self-reflexive and therefore socially caused; hence the strong thesis is false. But the argument from self-refutation is stronger than that, because it entails that the concept of what it is to be 'true' or 'false' must be changed to enable us to assert consistently that the strong thesis is true.

19 *Modes of Thought*, ed. R. Horton and R. Finnegan, pp. 237–8.

20 *Scientific Knowledge and Sociological Theory*, ch. 3.

21 See for example E. Durkheim, *The Elementary Forms of the Religious Life*, 1912; English edn, London, 1915, p. 423 f.

22 Cf. Bloor's and Caneva's use of Mannheim's suggestion that there are recurring correlations of romanticism with conservatism, and ration-

alism with liberalism (Bloor, *op. cit.*, p. 54; K. L. Caneva, 'From Galvanism to electrodynamics: the transformations of German physics and its social context', *Historical Studies in the Physical Studies*, ed. R. McCormmack, L. Pyenson, and R. S. Turner, vol. ix, 1978, 63; K. Mannheim, *Essays on Sociology and Social Psychology*, London, 1953, ch. 2. The 'archeology of ideas' of M. Foucault is an attempt to find isomorphisms of 'deep' structure between the different sciences within a given historical period; *The Order of Things*, English edn, London, 1970; *The Archaeology of Knowledge*, English edn, London, 1972. Mary Douglas' 'grid-group' hypothesis in the sociology of ideas finds correlations between degrees of symbolic elaboration of social life on the one hand, and the strength of individual-group authority structures on the other. This hypothesis has been applied to scientific cosmologies and their social provenance by Mary Douglas herself, *Natural Symbols*, London, 1970; *Implicit Meanings*, London, 1975, Part III; and *Cultural Bias,* London, 1978, and by historians of science including D. Bloor, K. L. Caneva and M. J. Rudwick *Exercises in Cultural Analysis: Grid-Group Analysis*, ed. M. Douglas and D. Ostrander, New York, 1979.

23 See, for example, J. Passmore, *Priestley's Writings on Philosophy, Science and Politics*, New York, 1965 and R. H. Kargon, *Atomism in England from Hariot to Newton*, Oxford, 1966.

24 D. A. Mackenzie and S. B. Barnes, 'Biometrician versus Mendelian: a controversy and its explanation', *Kölner Zeits. für Soziologie und Sozialpsychologie*, vol. xviii, 1975, 165, and *University of Edinburgh Science Studies Unit,* 1974.

25 *Ibid.*, p. 185. Barnes seems to have shifted somewhat between his first and second books towards a greater sympathy with what I have called 'intensive' methods, and a correspondingly weaker stress on the 'scientific' and even determinist character of sociology of science. Compare 'The determinist stance [with regard to human knowledge] can be seen as a natural extension of scientific theorizing: it is a characteristically scientific move to make' (*Scientific Knowledge and Sociological Theory*, p. 81), and 'no laws or necessary connections are proposed to link knowledge and the social order . . . the basic claim that knowledge is a resource in activity and not a direct determinant of it makes such an approach inappropriate' (*Interests and the Growth of Knowledge*, p. 85).

26 Cf. Weber's method of *Verstehen*, which was not intended as a species of subjective empathy, but rather as a mutual recognition of rationality by actors and social scientists. It follows that the method is only ideally applicable where the context is as near as possible to logic and empirical science. See, for example, *The Theory of Economic and Social Organization*, English edn, Oxford, 1947, ch. 1.

27 Cf. *The Protestant Ethic and the Spirit of Capitalism*, English edn, London, 1930; *The Sociology of Religion*, English edn, London, 1963, vols. xv, xvi.

28 See note 23.

II OBJECTIVITY AND TRUTH

3 *Theory and Observation*

I. Is there an independent observation language?

Rapidity of progress, or at least change, in the analysis of scientific theory structure is indicated by the fact that only a few years ago the natural question to ask would have been 'Is there an independent theoretical language?' The assumption would have been that theoretical language in science is parasitic upon observation language, and probably ought to be eliminated from scientific discourse by disinterpretation and formalization, or by explicit definition in or reduction to observation language. Now, however, several radical and fashionable views place the onus on believers in an observation language to show that such a concept has any sense in the absence of a theory. It is time to pause and ask what motivated the distinction between a so-called theoretical language and an observation language in the first place, and whether its retention is not now more confusing than enlightening.

In the light of the importance of the distinction in the literature, it is surprisingly difficult to find any clear statement of what the two languages are supposed to consist of. In the classic works of twentieth-century philosophy of science, most accounts of the observation language were dependent on circular definitions of observability and its cognates, and the theoretical language was generally defined negatively as consisting of those scientific terms which are not observational. We find quasi-definitions of the following kind: 'Observation-statement' designates a statement 'which records an actual or possible observation'; 'Experience, observation, and cognate terms will be used in the widest sense to cover observed facts about material objects or events in them as well as directly known facts about the contents or objects of immediate experience'; 'The observation language uses terms designating observable properties and relations for the description of observable things or events'; '*observables*, i.e., . . . things and events which are ascertainable by direct observation'.[1] Even Nagel, who gives the most thorough account of the alleged distinction between theoretical and observation terms, seems to presuppose that there is nothing problematic about the 'direct experimental evidence' for observation statements, or the 'experimentally identifiable instances' of observation terms.[2]

In contrast with the allegedly clear and distinct character of the observation terms, the meanings of theoretical terms, such as 'electron', 'electromagnetic wave' and 'wave function',[3] were held to be obscure. Philosophers have dealt with theoretical terms by various methods, based on the assumption that they have to be explained by means of the observation terms as given. None of the suggested methods has, however, been shown to leave theoretical discourse uncrippled in some area of its use in science. What suggests itself, therefore, is that the presuppositions of all these methods themselves are false, namely

(*a*) that the meanings of the observation terms are unproblematic;

(*b*) that the theoretical terms have to be understood by means of the observation terms; and

(*c*) that there is, in any important sense, a distinction between two *languages* here, rather than different kinds of uses within the same language.

In other words, the fact that we somehow understand, learn and use observation terms does not in the least imply that the way in which we understand, learn and use them is either different from or irrelevant to the way we understand, learn and use theoretical terms. Let us then subject the observation language to the same scrutiny which the theoretical language has received.

Rather than attacking directly the dual language view and its underlying empiricist assumptions, my strategy will be first to attempt to construct a different account of meaning and confirmation in the observation language. This project is not the ambitious one of a general theory of meaning, nor of the learning of language, but rather the modest one of finding conditions for understanding and use of terms in science—some specification, that is to say, in a limited area of discourse, of the 'rules of usage' which distinguish meaningful discourse from mere vocal reflexes. In developing this alternative account I shall rely on ideas which have become familiar particularly in connection with Quine's discussions of language and meaning and the replies of his critics, whose significances for the logic of science seem not yet to have been exploited nor even fully understood.

I shall consider in particular the predicate terms of the so-called observation language. But first something must be said to justify considering the problem as one of 'words' and not of 'sentences'. It has often been argued that it is sentences that we learn, produce, understand and respond to, rather than words; that is, that in theoretical discussion of language, sentences should be taken as units. There are, however, several reasons why this thesis, whether true or false, is irrelevant to the present problem, at least in its preliminary stages. The observation language of science is only a segment of the natural language in which it is expressed, and we may for the moment assume that rules of sentence formation and grammatical connectives are already given when we come to consider the use of observation predicates. Furthermore, since we are interested in alleged distinctions between the observation and theoretical languages, we are likely to find these distinctions in the characteristics of their respective predicates, not in the connectives which we may assume that they share. Finally, and most importantly, the present enterprise does not have the general positive aim of describing the entire structure of a language. It has rather the negative aim of showing that there are no terms in the observation language which are sufficiently accounted for by 'direct observation', 'experimentally identifiable instances' and the like. This can best be done by examining the hardest cases, that is, predicates which do appear to have direct empirical reference. No one would seriously put forward the direct-observation account of grammatical connectives; and if predicates are shown not to satisfy the account, it is likely that the same arguments will suffice to show that sentences do not satisfy it either.

So much for preliminaries. The thesis I am going to put forward can be briefly stated in two parts.

1. All descriptive predicates, including observation and

theoretical predicates, must be introduced, learned, understood and used, either by means of direct empirical associations in some physical situations, or by means of sentences containing other descriptive predicates which have already been so introduced, learned, understood and used, or by means of both together. (Introduction, learning, understanding and use of a word in a language will sometimes be summarized in what follows as the *function* of that word in the language.)

2. No predicates, not even those of the observation language, can function by means of direct empirical associations alone.

The process of functioning in the language can be spelled out in more detail.

A. Some predicates are initially learned in empirical situations in which an association is established between some aspects of the situation and a certain word. Given that any word with extralinguistic reference is ever learned, this is a necessary statement and does not presuppose any particular theory about what an association is or how it is established. This question is one for psychology or linguistics rather than philosophy. Two necessary remarks can, however, be made about such learning.

(1) Since every physical situation is indefinitely complex, the fact that the particular aspect to be associated with the word is identified out of a multiplicity of other aspects implies that degrees of physical similarity and difference can be recognized between different situations.

(2) Since every situation is in detail different from every other, the fact that the word can be correctly reused in a situation in which it was not learned has the same implication.

These remarks would seem to be necessarily implied in the premiss that some words with reference are learned by empirical associations. They have not gone unchallenged, however, and it is possible to distinguish two sorts of objections to them. First, some writers, following Wittgenstein, have appeared to deny that physical similarity is necessary to the functioning of *any* word with extralinguistic reference. That similarity is not *sufficient*, I am about to argue, and I also agree that not all referring words need to be introduced in this way, but if *none* were, I am unable to conceive how an intersubjective descriptive language could ever get under way. The onus appears to rest upon those who reject similarity to show in what other way descriptive language is possible. For example, Donald Davidson claims that there is no need for a descriptive predicate to be learned in the presence of the object to which it is properly applied, since, for example, it might be learned in 'a skilfully faked environment'.[4] This possibility does not, however, constitute an objection to the thesis that it must be learned in *some* empirical situation, and that this situation must have some similarity with those situations in which the predicate is properly used. Chomsky, on the other hand, attacks what he regards as Quine's 'Humean theory' of language acquisition by resemblance of stimuli, and conditioned response.[5] But the necessity of the *similarity* condition for language learning does not depend on the particular empirical mechanism of learning. Learning by patterning the environment in terms of a set of 'innate ideas' would depend equally upon subsequent application of the same pattern to similar features of the environment. Moreover, 'similar' cannot just be *defined* as 'properly ascribed the same descriptive predicate in the same language community', since for one thing similarity is a matter of degree and is a non-transitive relation, whereas 'properly ascribed the same descriptive predicate' is not, or not obviously. The two terms can therefore hardly be synonymous. I therefore take it as a necessary *a priori* condition of the applicability of a language containing universal terms that *some* of these terms presuppose primitive causal recognitions of physical similarities.

A different sort of objection to the appeal to similarity is made by Popper, who argues that the notion of repetition of instances which is implied by (1) and (2) is essentially vacuous, because similarity is always similarity *in certain respects*, and 'with a little ingenuity' we could always find similarities in *some* same respects between all members of any finite set of situations. That is to say, 'anything can be said to be a "repetition" of anything, if only we adopt the appropriate point of view'.[6] But if this were true, it would make the learning process in empirical situations impossible. It would mean that however finitely large the number of presentations of a given situation-aspect, that aspect could never be identified as the desired one out of the indefinite number of other respects in which the presented situations are all similar. It would, of course, be possible to eliminate some other similarities by presenting further situations similar in the desired respect but not in others, but it would then be possible to find other respects in which all the situations, new and old, are similar—and so on without end.

However, Popper's admission that 'a little ingenuity' may be required allows a less extreme interpretation of his argument, namely that the physics and physiology of situations already give us some 'point of view' with respect to which some pairs of situations are similar in more obvious respects than others, and one situation is more similar in some respect to another than it is in the same respect to a third. This is all that is required by the assertions (1) and (2). Popper has needlessly obscured the importance of these implications of the learning process by speaking as though, before any repetition can be recognized, we have to take thought and *explicitly* adopt a point of view. If this were so, a regressive problem would arise about how we ever learn to apply the predicates in which we explicitly express that point of view. An immediate consequence of this is that there must be a stock of predicates in any descriptive language for which it is impossible to *specify* necessary and sufficient conditions of correct application. For if any such specification could be given for a particular predicate, it would introduce further predicates requiring to be learned in empirical situations for which there was no specification. Indeed, such unspecified predicates would be expected to be in the majority, for those for which necessary and sufficient conditions can be given are dispensable except as a shorthand and hence essentially uninteresting. We must therefore conclude that the primary process of recognition of similarities and differences is necessarily *unverbalizable*. The emphasis here is of course on *primary*, because it may be perfectly possible to give empirical descriptions of the conditions, both psychological and physical, under which similarities are recognized, but such descriptions will themselves depend on further undescribable primary recognitions.

B. It may be thought that the primary process of classifying objects according to recognizable similarities and differences will provide us with exactly the independent observation predicates required by the traditional view. This, however, is to overlook a logical feature of relations of similarity and difference, namely that they are not *transitive*. Two objects *a* and *b* may be judged to be similar to some degree in respect to predicate *P*, and may be placed in the class of objects to which *P* is applicable. But object *c* which is judged similar to *b* to the same degree may not be similar to *a* to the same degree or indeed to any degree. Think of judgments of similarity of three shades of colour. This leads to the conception of some objects as being more 'central' to the *P*-class than others, and also implies that the process of classifying objects by recognition of similarities and differences is necessarily accompanied by some loss of (unverbalizable) information. For if *P* is a predicate whose conditions of applicability are dependent on the process just described, it is impossible to *specify* the degree to which an object satisfies *P* without introducing more predicates about which the same story would have to be told. Somewhere this potential regress must be stopped by some predicates whose application involves loss of information which is present to recognition but not verbalizable. However, as we shall see shortly, the primary recognition process, though necessary, is not sufficient for classification of objects as *P*, and the loss of information involved in classifying leaves room for changes in classification to take place under some circumstances. Hence primary recognitions do not provide a stable and independent list of primitive observation predicates.

C. It is likely that the examples that sprang to mind during the reading of section *B* were such predicates as 'red', 'ball' and 'teddy bear'. But notice that nothing that has been said rules out the possibility of giving the same account of apparently much more complex words. 'Chair', 'dinner' and 'mama' are early learned by this method, and it is not inconceivable that it could also be employed in first introducing 'situation', 'rule', 'game', 'stomach ache' and even 'heartache'. This is not to say, of course, that complete fluency in using these words could be obtained by this method alone; indeed, I am now going to argue that complete fluency cannot be obtained in the use of *any* descriptive predicate by this method alone. It should only be noticed here that it is possible for any word in natural language having some extralinguistic reference to be introduced in suitable circumstances in some such way as described in section *A*.

D. As learning of the language proceeds, it is found that some of these predicates enter into the general statements which are accepted as true and which we will call *laws*: 'Balls are round'; 'In summer leaves are green'; 'Eating unripe apples leads to stomach ache'. It matters little whether some of these are what we would later come to call analytic statements; some, perhaps most, are synthetic. It is not necessary, either, that every such law should be *in fact* true, only that it is for the time being accepted as true by the language community. As we shall see later, any one of these laws may be *false* (although not all could be false at once). Making explicit these general laws is only a continuation and extension of the process already described as identifying and reidentifying proper occasions of the use of a predicate by means of physical similarity. For knowledge of the laws will now enable the language user to apply descriptions correctly in situations other than those in which he learned them, and even in situations where nobody could have learned them in the absence of the laws—for example, 'stomach ache' of an absent individual known to have consumed a basketful of unripe apples, or even 'composed of diatomic molecules' of the oxygen in the atmosphere. In other words, the laws enable generally correct inferences and predictions to be made about distant ('unobservable') states of affairs.

E. At this point the system of predicates and their relations in laws has become sufficiently complex to allow for the possibility of internal misfits and even contradictions. This possibility arises in various ways. It may happen that some of the applications of a word in situations turn out not to satisfy the laws which are true of other applications of the word. In such a case, since degrees of physical similarity are not transitive, a reclassification may take place in which a particular law is preserved in a subclass more closely related by similarity, at the expense of the full range of situations of application which are relatively less similar. An example of this would be the application of the word 'element' to water, which becomes incorrect in order to preserve the truth of a system of laws regarding 'element', namely that elements cannot be chemically dissociated into parts which are themselves elements, that elements always enter into compounds, that every substance is constituted by one or more elements, and so on. On the other hand, the range of applications may be widened in conformity with a law, so that a previously incorrect application becomes correct. For example, 'mammal' is correctly applied to whales, whereas it was previously thought that 'Mammals live only on land' was a well-entrenched law providing criteria for correct use of 'mammal'. In such a case it is not adequate to counter with the suggestion that the correct use of 'mammal' is *defined* in terms of animals which suckle their young, for it is conceivable that if other empirical facts had been different, the classification in terms of habitat would have been more useful and comprehensive than that in terms of milk production. And in regard to the first example, it cannot be maintained that it is the *defining* characteristics of 'element' that are preserved at the expense of its application to water because, of the conditions mentioned, it is not clear that any particular one is, or ever has been, taken as *the* defining characteristic; and since the various characteristics are logically independent, it is empirically possible that some might be satisfied and not others. *Which* is preserved will always depend on what system of laws is most convenient, most coherent and most comprehensive. But the most telling objection to the suggestion that correct application is decided by definition is of course the general point made at the end of section *A* that there is always a large number of predicates for which *no* definition in terms of necessary and sufficient conditions of application can be given. For these predicates it is possible that the primary recognition of, for example, a whale as being sufficiently similar to some fish to justify its inclusion in the class of fish may be explicitly overridden in the interests of preserving a particular set of laws.

Properly understood, the point developed in the last paragraph should lead to a far-reaching reappraisal of orthodoxy regarding the theory–observation distinction. To summarize, it entails that no feature in the total landscape of functioning of a descriptive predicate is exempt from modification under pressure from its surroundings. That any empirical law may be abandoned in the face of counter-examples is trite, but it becomes less trite when the functioning of every predicate is found to depend essentially on some laws or other and when it is also the case that any 'correct' situation of application—*even that in terms of which the term was originally introduced*—may become incorrect in order to preserve a system of laws and other applications. It is in this sense that I shall understand the 'theory dependence' or 'theory-ladenness' of all descriptive predicates.

One possible objection to this account is easily anticipated. It is not a *conventionalist* account, if by that we mean that any law can be assured of truth by sufficiently meddling with the meanings of its predicates. Such a view does not take seriously the systematic character of laws, for it contemplates preservation of the truth of a given law irrespective of its coherence with the rest of the system, that is, the preservation of simplicity and other desirable internal characteristics of the system. Nor does it take account of the fact that not all primary recognitions of empirical similarity can be overridden in the interest of preserving a given law, for it is upon the existence of some such recognitions that the whole possibility of language with empirical reference rests. The present account on the other hand demands both that laws shall remain connected in an economical and convenient system and that at least most of its predicates shall remain applicable, that is, that they shall continue to depend for applicability upon the primary recognitions of similarity and difference in terms of which they were learned. That it is possible to have such a system with a given set of laws and predicates is not a convention but a fact of the empirical world. And although this account allows that *any* of the situations of correct application may change, it cannot allow that *all* should change, at least not all at once. Perhaps it would even be true to say that only a small proportion of them can change at any one time, although it is conceivable that over long periods of time most or all of them might come to change piecemeal. It is likely that almost all the terms used by the alchemists that are still in use have now changed their situations of correct use quite radically, even though at any one time chemists were preserving most of them while modifying others.

II. Entrenchment

It is now necessary to attack explicitly the most important and controversial question in this area, namely the question whether the account of predicates that has been given really applies to all descriptive predicates whatsoever, or whether there are after all some that are immune to modification in the light of further knowledge and that might provide candidates for a basic and independent observation language. The example mentioned at the end of the last paragraph immediately prompts the suggestion that it would be possible at any time for both alchemists and chemists to 'withdraw' to a more basic observation language than that used in classifying substances and that this language would be truly primitive and theory-independent. The suspicion that this may be so is not even incompatible with most of the foregoing account, for it may be accepted that we often do make words function without reflecting upon more basic predicates to which we could withdraw if challenged. Thus, it may not be disputed that we learn, understand and use words like 'earth', 'water', 'air' and 'fire' in empirical situations and that their subsequent functioning depends essentially upon acceptance of some laws; and yet it may still be maintained that there are some more basic predicates for which cash value is payable in terms of empirical situations alone. Let us therefore consider this argument at its strongest point and take the case of the putative observation predicate 'red'. Is this predicate subject to changes of correct application in the light of laws in the way that has been described? The defence of our account at this point comes in two stages. First, it must be shown that *no* predicate of an observation language can function by mere empirical situations alone, independently of any laws. Second, it must be shown that there is no set of observation predicates whose interrelating laws are absolutely invariant to changes in the rest of the network of laws.

When a predicate such as 'red' is claimed to be 'directly' descriptive, this claim is usually made in virtue of its use as a predicate of immediate experience—a sensation of a red postage stamp, a red spectral line, a red after-image. It is unnecessary here to enter into the much discussed questions of whether there are any such 'things' as sensations for 'red' to be a predicate of, whether such predicates of sensations could be ingredients of either a public or a private language, and whether there is indeed any sense in the notion of a of a private language. The scientific observation language at least is not private but must be intersubjective; and whether some of its predicates are predicates of sensations or not, it is still possible to raise the further question: in *any* intersubjective language can the functioning of the predicates be independent of accepted laws? That the answer is negative can be seen by considering the original account of the empirical situations given in section I.*A* and by adopting one generally acceptable assumption. The assumption is that in using a public language, the correctness of any application of a predicate in a given situation must in principle be capable of intersubjective test.[7] Now if my careful response of 'red' to each of a set of situations were all that were involved in my correct use of 'red', this response would not be sufficient to ensure intersubjectivity. It is possible, in spite of my care, that I have responded mistakenly, in which case the laws relating 'red' to other predicates can be appealed to in order to correct me (I can even correct myself by this method): 'It can't have been red, because it was a sodium flame, and sodium flares are not red.' If my response 'red' is intended to be an ingredient of a public observation language, it carries at least the implication that disagreements can be publicly resolved, and this presupposes laws conditioning the function of 'red'. If this implication is absent, responses are mere verbal reflexes having no intersubjective significance (unless of course they are part of a physiological–psychological experiment, but then I am subject, not observer). This argument does not, I repeat, purport to show that there could not be sense-datum language functioning as the observation language of science—only that if this were so, its predicates would share the double aspect of empirical situation and dependence on law which belongs to all putative observation predicates.

Now consider the second stage of defence of our account. The suggestion to be countered here is that even if there are peripheral uses of 'red' which might be subject to change in the event of further information about laws, there is nevertheless a central core of function of 'red', with at least some laws which ensure its intersubjectivity, which remains stable throughout all extensions and modifications of the rest of the network of accepted laws. To illustrate the contrast between 'periphery' and 'core' take the following examples: we might come to realize that when 'red' is applied to a portion of the rainbow, it is not a predicate of an object, as in the paradigm cases of 'red', or that the ruddy hue of a distant star is not the colour of the star but an effect of its recession. But, it will be said, in regard to cherries, red lips and the colour of a strontium compound in a Bunsen flame, 'red' is used entirely independently of the truth of or knowledge of the great majority of laws in our network. We might of course be mistaken in application of 'red' to situations of this central kind, for we may mistake colour in a bad light or from defects of vision; but there are enough laws whose truth cannot be in doubt to enable us to correct mistakes of this kind, and by appealing to them we are always able to come to agreement about correct applications. There is no sense, it will be argued, in supposing that in cases like this we could all be mistaken all the time or that we might, in any but the trivial sense of deciding to use another word equivalent to 'red', come to change our usage in these central situations.

One possible reply[8] is to point out that the admission that there are *some* situations in which we might change our use even of a predicate like 'red' is already a significant one, especially in the examples given above. For the admission that the 'red' of a rainbow or a receding star is not the colour of an object is the admission that in these cases at least it is a *relational* predicate, where the relata, which may be quite complex, are spelled out by the laws of physics. Now no doubt it does not *follow* that 'red' ascribed to the book cover now before me is also a relational predicate, unless we take physics to provide the real truth about everyday objects as well as those that are more remote. The schizophrenia induced by not taking physics seriously in this way raises problems of its own which we cannot pursue here. But suppose our critic accepts the realist implication that 'red' is, on all occasions of its use as a predicate of objects, in fact a relational predicate, and then goes on to discount this admission by holding that such a relatively subtle logical point is irrelevant to the ordinary function of 'red' in the public language. Here we come near the heart of what is true in the critic's view. The truth might be put like this: Tom, Dick and Mary do indeed use the word 'red' with general indifference to logical distinctions between properties and relations. Even logicians and physicists continue to use it in such a way that in ordinary conversation it need never become apparent to others, or even to themselves, that they 'really believe' that colour predicates are relational. And more significantly for the ultimate purpose of this essay, the conversation of a Newtonian optician about sticks and stones and rolls of bread need never reveal a difference of function of 'red' from the conversation of a post-relativity physicist.

Such a concession to the critic with regard to invariance of function in limited domains of discourse is an important one, but it should be noticed that its force depends not upon fixed stipulations regarding the use of 'red' in particular empirical situations, but rather upon empirical facts about the way the world is. Physically possible situations can easily be envisaged in which even this central core of applicability of 'red' would be broken. Suppose an isolated tribe all suffered a congenital colour blindness which resulted in light green being indistinguishable from red and dark green from black. Communication with the outside world, or even the learning of physics without any such communication, might very well lead them to revise the function of 'red and 'black' even in paradigm cases of their own language.

A more realistic and telling example is provided by the abandonment of Newtonian time simultaneity. This is an especially striking case, because time concepts are among the most stable in most languages and particularly in a physics which has persistently regarded spatial and temporal qualities as primary and as providing the indispensable framework of a mechanistic science. As late as 1920 N. R. Campbell, usually a perceptive analyst of physical concepts, wrote:

> Is it possible to find any judgment of sensation concerning which all sentient beings whose opinion can be ascertained are always and absolutely in agreement? . . . I believe that it is possible to obtain absolutely universal agreement for judgments such as, the event A happened at the same time as B, or A happened between B and C.[9]

Special relativity had already in 1905 shown this assumption to be false. This means that at any time before 1905 the assumption was one from which it was certainly possible to withdraw; it was in fact 'theory-laden', although it had not occurred to anybody that this was the case. Now let us cast Einstein in the role of the 'operationist' physicist who, wiser than his contemporaries, has detected the theory-ladenness and wishes to withdraw from it to a 'level of direct observation', where there are no theoretical implications, or at least where these are at a minimum.[10] What can he do? He can try to set up an operational definition of time simultaneity. When observers are at a distance from each other (they are alway at *some* distance), and when they are perhaps also moving relatively to each other, he cannot assume that they will agree on judgments of simultaneity. He will assume only that a given observer can judge events that are simultaneous in his own field of vision, provided they occur close together in that field. The rest of Einstein's operational definition in terms of light signals between observers at different points is well known. But notice that this definition does not carry out the programme just proposed for an operationist physicist. For far from withdrawal to a level of direct observation where theoretical implications are absent or at a minimum, the definition requires us to assume, indeed to postulate, that the velocity of light *in vacuo* is the same in all directions and invariant to the motions of source and receiver. This is a postulate which is logically prior in special relativity to any experimental measurement of the velocity of light, because it is used in the very definition of the time scale at distant points. But from the point of view of the operationist physicist before 1905, the suggestion of withdrawing from the assumption of distant absolute time simultaneity to this assumption about the velocity of light could not have appeared to be a withdrawal to more direct observation having fewer theoretical implications but rather the reverse. This example illustrates well the impossibility of even talking sensibly about 'levels of more direct observation' and 'degrees of theory-ladenness' *except in the context of some framework of accepted laws*. That such talk is dependent on this context is enough to refute the thesis that the contrast between 'direct observation' and 'theory-ladenness' is itself theory-independent. The example also illustrates the fact that at any given stage of science it is never possible to know *which* of the currently entrenched predicates and laws may have to give way in the future.

The operationist has a possible comeback to this example. He may suggest that the process of withdrawal to the directly observed is not a process of constructing another theory, as Einstein did, but properly stops short at the point where we admitted that at least one assumption of Newtonian physics is true and must be retained, namely, that 'a given observer can judge events that are simultaneous in his own field of vision, provided they occur close together in that field—call this assumption *S*. This, it may be said, is a genuine withdrawal to a less theory-laden position, and all that the rest of the example shows is that there is in fact no possibility of advance again to a more general conception of time simultaneity without multiplying insecure theoretical assumptions. Now, of course, the game of isolating some features of an example as paradigms of 'direct observation', and issuing a challenge to show how *these* could ever be overthrown, is one that can go on regressively without obvious profit to either side. But such a regress ought to stop if either of the following two conditions is met:

(a) that it is logically possible for the alleged paradigm to be overthrown and that its overthrow involves a *widening* circle of theoretical implications;
or
(b) that the paradigm becomes less and less suitable as an observation statement, because it ceases to have the required intersubjective character.

The time simultaneity example made its point by illustrating condition (a). The assumption *S* to which it is now suggested we withdraw can be impaled on a dilemma between (a) and (b). Suppose it were shown that an observer's judgment of simultaneity in his field of sensation were quite strongly dependent on the strength of the gravitational field in his neighbourhood, although this dependence had not yet been shown up in the fairly uniform conditions of observation on the surface of the earth. Such a discovery, which is certainly conceivable, would satisfy condition (a). As long as the notion of simultaneity is so interpreted as to allow for intersubjective checking and agreement, there is always an indefinite number of such possible empirical situations whose variation might render the assumption *S* untenable. The only way to escape this horn of the dilemma is to interpret *S* as referring to the direct experience of simultaneity of a single observer, and this is intersubjectively and hence scientifically useless, and impales us on the horn of condition (b).

The comparative stability of function of the so-called observation predicates is logically speaking an accident of the way the world is. But it may now be suggested that since the way the world is is not likely to alter radically during the lifetime of any extant language, we might define an observation language to be just that part of language which the facts allow to remain stable. This, however, is to take less than seriously the effects of scientific knowledge on our ways of talking about the world and also to underestimate the tasks that ordinary language might be called upon to perform as the corpus of scientific knowledge changes. One might as well hold that the ordinary language of Homer, which identifies life with the breath in the body and fortuitous events with interventions of divine personages, and was no doubt adequate to discourse before the walls of Troy, should have remained stable in spite of all subsequent changes in physics, physiology, psychology and theology. Our ordinary language rules for the use of 'same time', which presuppose that this concept is independent of the distance and relative motion of the spatial points at which time is measured, are not only contradicted by relativity theory, but would possibly need fundamental modification if we all were to take habitually to space travel. Another point to notice here is that the comparatively stable area within which it is proposed to define an observation language itself is partly known to us because its stability is explained by the theories we now accept. It is certainly not sufficiently defined by investigating what observation statements have in fact remained stable during long periods of time, for this stability might be due to accident, prejudice or false beliefs. Thus any attempted definition itself would rely upon current theories and hence not be a definition of an observation language which is theory-independent. Indeed, it might justly be concluded that we shall know what the most adequate observation language is only when, if possible, we have true and complete theories, including theories of physiology and physics which tell us what it is that is most 'directly observed'. Only then shall we be in a position to make the empirical distinctions that seem to be presupposed by attempts to discriminate theoretical and observation predicates.

The upshot of all this may be summarized by saying that although there is a nucleus of truth in the thesis of invariance of the observation language and, hence, of the theory–observation distinction among predicates, this truth has often been located in the wrong place and used to justify the wrong inferences. The invariance of observation predicates has been expressed in various ways, not all equivalent to one another and not all equally valid.[11] Let us summarize the discussion so far by examining some of these expressions.

(i) 'There are some predicates that are *better entrenched* than others, for instance, "red" than "ultra-violet", "lead" than "π-meson".'

If by 'better entrenched' is meant less subject to change of function in ordinary discourse and therefore less revelatory of the speaker's commitments to a system of laws or of his relative ignorance of such systems, then (i) is true. But this is a *factual* truth about the relative invariance of some empirical laws to increasing empirical information, not about the *a priori* features of a peculiar set of predicates, and it does not entail that any predicate is *absolutely* enriched, nor that any subsystems of predicates and the laws relating them are immune to modification under pressure from the rest of the system.

(ii) 'There are some predicates that refer to aspects of situations more *directly observable* than others.'

If this means that their function is more obviously related to empirical situations than to laws, (ii) is true, but its truth does not imply that a line can be drawn between theoretical and observation predicates in the place it is usually desired to draw it. For it is not at all clear that highly complex and even theoretical predicates may not sometimes be directly applicable in appropriate situations. Some examples were given in section I.*C*; other examples are thinkable where highly theoretical descriptions would be given directly: 'particle-pair annihilation' in a cloud chamber, 'glaciation' of a certain landscape formation, 'heart condition' of a man seen walking along the street. To the immediate rejoinder that these examples leave open the possibility of withdrawal to less 'theory-laden' descriptions, a reply will be given in (v) below. Meanwhile it should be noticed that this sense of 'observable' is certainly not coextensive with that of (i).

(iii) 'There are some predicates that are learnable and applicable in a *pragmatically* simpler and quicker manner than others.'

This is true, but does not necessarily single out the same set of predicates in all language communities. Moreover, it does not necessarily single out all or only the predicates that are 'observable' in senses (i) and (ii).

(iv) 'There are some predicates in terms of which others are *anchored to the empirical facts*.'

This may be true in particular formulations of a theory, where the set of anchor predicates is understood as in (i), (ii) or (iii), but little more needs to be said to justify the conclusion that such a formulation and its set of anchoring predicates would not be unique. In principle it is conceivable that any predicate could be used as a member of the set. Thus, the commonly held stronger version of this assumption is certainly false, namely that the anchor predicates have unique properties which allow them to endow theoretical predicates with empirical meaning which these latter would not otherwise possess.

(v) The most important assumption about the theory–observation distinction, and the one which is apparently most damaging to the present account, can be put in a weaker and a stronger form:

(a) 'There are some predicates to which we could always *withdraw* if challenged in our application of others.'

(b) 'These form a unique subset in terms of which "pure descriptions" free from "theory-loading" can be given.'

Assumption (a) must be accepted to just the extent that we have accepted the assumption that there are degrees of entrenchment of predicates, and for the same reasons. It is indeed sometimes possible to withdraw from the implications of some ascriptions of predicates by using others better entrenched in the network of laws. To use some of the examples already mentioned, we may withdraw from 'particle-pair annihilation' to 'two white streaks meeting and terminating at an angle'; from 'heart condition' to a carefully detailed report of complexion, facial structure, walking habits and the like; and from 'epileptic fit' to a description of teeth-clenching, falling, writhing on the floor and so on. So far, these examples show only that some of the lawlike implications that are in mind when the first members of each of these pairs of descriptions are used can be withdrawn from and replaced by descriptions which do not have *these* implications. They do not show that the second members of each pair are free from lawlike implications of their own, nor even that it is possible to execute a series of withdrawals in such a way that each successive description contains fewer implications than the description preceding it. Far less do they show that there is a unique set of descriptions which have *no* implications; indeed the arguments already put forward should be enough to show that this assumption, assumption (b), must be rejected. As in the case of entrenchment, it is in principle possible for any particular lawlike implication to be withdrawn from, although not all can be withdrawn from at once. Furthermore, although in any given stage of the language some descriptive predicates are more entrenched than others, it is not clear that withdrawal to those that are better entrenched is withdrawal to predicates which have *fewer* lawlike implications. Indeed, it is likely that better entrenched predicates have in fact far more implications. The reason why these implications do not usually seem doubtful or objectionable to the observational purist is that they have for so long proved to be true, or been believed to be true, in their relevant domains that their essentially inductive character has been forgotten. It follows that when well-entrenched predicates and their implications are from time to time abandoned under pressure from the rest of the network, the effects of such abandonment will be more far-reaching, disturbing and shocking than when less well-entrenched predicates are modified.

III. The network model

The foregoing account of theories, which has been presented as more adequate than the deductive two-language model, may be dubbed the *network model* of theories. It is an account that was first explicit in Duhem and more recently reinforced by Quine. Neither in Duhem nor in Quine, however, is it quite clear that the netlike interrelations between more directly observable predicates and their laws are in principle just as subject to modifications from the rest of the network as are those that are relatively theoretical. Duhem seems sometimes to imply that although there is a network of relatively phenomenological representations of facts, once established this network remains stable with respect to the changing explanations. This is indeed one reason why he rejects the view that science aims at explanation in terms of unobservable entities and restricts theorizing to the articulation of mathematical representations which merely systematize but do not explain the facts. At the same time, however, his analysis of the facts is far subtler than that presupposed by later deductivists and instrumentalists. He sees that what is primarily significant for science is not the precise nature of what we directly observe, which in the end is a *causal* process, itself susceptible of scientific analysis. What is significant is the interpretative expression we give to what is observed, what he calls the *theoretical facts*, as opposed to the 'raw data' represented by *practical facts*. This distinction may best be explained by means of his own example. Consider the theoretical fact 'The temperature is distributed in a certain manner over a certain body'.[12] This, says Duhem, is susceptible of precise mathematical formulation with regard to the geometry of the body and the numerical specification of the temperature distribution. Contrast the practical fact. Here geometrical description is at best an idealization of a more or less rigid body with a more or less indefinite surface. The temperature at a given point cannot be exactly fixed, but is only given as an average value over vaguely defined small volumes. The theoretical fact is in imperfect translation, or interpretation, of the practical fact. Moreover, the relation between them is not one-to-one but rather many-to-many, for an infinity of idealizations may be made to more or less fit the practical fact, and an infinity of practical facts may be expressed by means of one theoretical fact.

Duhem is not careful in his exposition to distinguish *facts* from *linguistic expressions of facts*. Sometimes both practical and theoretical facts seem to be intended as linguistic statements (for instance, where the metaphor of 'translation' is said to be appropriate). But even if this is his intention, it is clear that he does not wish to follow traditional empiricism into a search for forms of expression of practical facts which will constitute the basis of science. Practical facts are not the appropriate place to look for such a basis—they are imprecise, ambiguous, corrigible, and on their own ultimately meaningless. Moreover, there is a sense in which they are literally inexpressible. The absence of distinction between fact and linguistic expression here is not accidental. As soon as we begin to try to capture a practical fact in language, we are committed to some theoretical interpretation. Even to say of the solid body that 'its points are more or less worn down and blunt' is to commit ourselves to the categories of an ideal geometry.

What, then, is the 'basis' of scientific knowledge for Duhem? If we are to use this conception at all, we must say that the basis of science is the set of theoretical facts in terms of which experience is interpreted. But we have just seen that theoretical facts have only a more or less loose and ambiguous relation with experience. How can we be sure that they provide a firm empirical foundation? The answer must be that we cannot be sure. There is no such foundation. Duhem himself is not consistent on this point, for he sometimes speaks of the persistence of the network of theoretical facts as if this, once established, takes on the privileged character ascribed to observation statements in classical positivism. But this is not the view that emerges from his more careful discussion of examples. For he is quite clear, as in the case of the correction of the 'observational' laws of Kepler by Newton's theory, that more comprehensive mathematical representations may show particular theoretical facts to be false.

However, we certainly seem to have a problem here, because if it is admitted that subsets of the theoretical facts may be removed from the corpus of science, and if we yet want to retain some form of empiricism, the decision to remove them can be made only be reference to *other* theoretical facts, whose status is in principle equally insecure. In the traditional language of epistemology some element of correspondence with experience, though loose and corrigible, must be retained but also be supplemented by a theory of the coherence of a network. Duhem's account of this coherence has been much discussed but not always in the context of his complete account of theoretical and practical facts, with the result that it has often been trivialized. Theoretical facts do not stand on their own but are bound together in a network of laws which constitutes the total mathematical representation of experience. The putative theoretical fact that was Kepler's third law of planetary motion, for example, does not fit the network of laws established by Newton's theory. It is therefore modified, and this modification is possible without violating experience because of the many-to-one relation between the theoretical fact and that practical fact understood as the ultimately inexpressible situation which obtains in regard to the orbits of planets.

It would seem to follow from this (although Duhem never explicitly draws the conclusion) that there is no theoretical fact or lawlike relation whose truth or falsity can be determined in isolation from the rest of the network. Moreover, many conflicting networks may more or less fit the same facts, and which one is adopted must depend on criteria other than the facts: criteria involving simplicity, coherence with other parts of science, and so on. Quine, as is well known, has drawn this conclusion explicitly in the strong sense of claiming that any statement can be maintained true in the fact of any evidence: 'Any statement can be held true come what may, if we make drastic enough adjustments elsewhere in the system. . . . Conversely, by the same token, no statement is immune to revision.'[13] In a later work, however, he does refer to 'the philosophical doctrine of infallibility of observation sentences' as being sustained in his theory. Defining the stimulus meaning of a sentence as the class of sensory stimulations that would prompt assent to the sentence, he regards observation sentences as those sentences whose stimulus meanings remain invariant to changes in the rest of the network and for which 'their stimulus meanings may without fear of contradiction be said to do full justice to their meanings'.[14] This seems far too conservative a conclusion to draw from the rest of the analysis, for in the light of the arguments and examples I have presented it appears very dubious whether there are such invariant sentences if a long enough historical perspective is taken.

There are other occasions on which Quine seems to obscure unnecessarily the radical character of his own position by conceding too much to more traditional accounts. He compares his own description of theories to those of Braithwaite, Carnap and Hempel in respect of the 'contextual definition' of theoretical terms. But his own account of these terms as deriving their meaning from an essentially *linguistic* network has little in common with the formalist notion of 'implicit definition' which these deductivists borrow from mathematical postulate systems in which the terms need not be interpreted empirically. In this sense the implicit definition of 'point' in a system of Riemannian geometry is entirely specified by the formal postulates of the geometry and does not depend at all on what would count empirically as a realization of such a geometry.[15] Again, Quine refers particularly to a net analogy which Hempel adopts in describing theoretical predicates as the knots in the net, related by definitions and theorems represented by threads. But Hempel goes on to assert that the whole 'floats . . . above the plane of observation' to which it is anchored by *threads of a different kind*, called 'rules of interpretation', *which are not part of the network itself.*[16] The contrast between this orthodox deductivism and Quine's account could hardly be more clear. For Quine, and in the account I have given here, there is indeed a network of predicates and their lawlike relations, but it is not floating above the domain of observation; it is attached to it at some of its knots. *Which* knots will depend on the historical state of the theory and its language and also on the way in which it is formulated, and the knots are not immune to change as science develops. It follows, of course, that 'rules of interpretation' disappear from this picture: *all* relations become laws in the sense defined above, which, it must be remembered, includes near analytic definitions and conventions as well as empirical laws.

IV. Theoretical predicates

So far it has been argued that it is a mistake to regard the distinction between theoretical and observational predicates either as providing a unique partition of descriptive predicates into two sets or as providing a simple ordering such that it is always possible to say of two predicates that one is under all circumstances more observational than or equally observational with the other. Various relative and non-coincident distinctions between theoretical and observational have been made, none of which is consistent with the belief that there is a unique and privileged set of observation predicates in terms of which theories are related to the empirical world. So far in the network model it has been assumed that any predicate may be more or less directly ascribed to the world in some circumstances or other, and that none is able to function in the language by means of such direct ascription alone. The second of these assumptions has been sufficiently argued; it is now necessary to say more about the first. Are there any descriptive predicates in science which could not under any circumstances be directly ascribed to objects? If there are, they will not fit the network model as so far described, for there will be nothing corresponding to the process of classification by empirical associations, even when this process is admitted to be fallible and subject to correction by laws, and they will not be connected to other predicates by laws, since a law presupposes that the predicates it connects have all been observed to co-occur in some situation or other.

First, it is necessary to make a distinction between theoretical *predicates* and theoretical *entities*, a distinction which has not been sufficiently considered in the deductivist literature. Theoretical entities have sometimes been taken to be equivalent to unobservable entities. What does this mean? If an entity is unobservable in the sense that it never appears as the subject of observation reports, and is not in any other way related to the entities which do appear in such reports, then it has no place in science. This cannot be what is meant by 'theoretical' when it is applied to such entities as electrons, mesons, genes and the like. Such applications of the terms 'theoretical' and 'unobservable' seem rather to imply that the entities do not have predicates ascribed to them in observation statements, but only in theoretical statements. Suppose the planet Neptune had turned out to be wholly transparent to all electro-magnetic radiation and, therefore, invisible. It might still have entered planetary theory as a theoretical entity in virtue of the postulated force relations between it and other planets. Furthermore, the monadic predicate 'mass' could have been inferred of it, although mass was never ascribed to it in an observation statement. Similarly, protons, photons and mesons have monadic and relational predicates ascribed to them in theoretical but not in observation statements, at least not in prescientific language. But this distinction, like others between the theoretical and the observational domains, is relative; for once a theory is accepted and further experimental evidence obtained for it, predicates may well be ascribed directly to previously unobservable entities, as when genes are identified with DNA molecules visible in micrographs or when the ratio of mass to charge of an elementary particle is 'read off' the geometry of its tracks in a magnetic field.

In contrasting theoretical with observable entities, I shall consider that theoretical entities are sufficiently specified as being those to which monadic predicates are not ascribed in relatively observational statements. It follows from this specification that relational predicates cannot be ascribed to them in observation statements either, for in order to recognize that a relation holds between two or more objects, it is necessary to recognize the objects by means of at least some monadic properties. ('The tree is to the left of *x*' is not an observation statement; 'the tree is to the left of *x* and *x* is nine storeys high' may be.) A theoretical entity must, however, have some postulated relation with any observable entity in order to enter scientific theory at all, and both monadic and relational predicates may be postulated of it in the context of a theoretical network. It must be emphasized that this specification is not intended as a close analysis of what deductivists have meant by 'theoretical entity' (which is in any case far from clear), but rather as an explication of this notion in terms of the network account of theories. At least it can be said that the typical problems that have seemed to arise about the existence of and reference to theoretical entities have arisen only in so far as these entities are not subjects of monadic predicates in observation statements. If a monadic predicate were ascribed to some entity in an observation statement it would be difficult to understand what would be meant by calling such an entity 'unobservable' or by questioning its 'existence'. The suggested explication of 'theoretical entity' is, therefore, not far from the apparent intentions of those who have used this term, and it does discriminate electrons, mesons and genes on the one hand from sticks and stones on the other.

When considering the relatively direct or indirect ascription of predicates to objects, it has already been argued that the circumstances of use must be attended to before the term 'unobservable' is applied. In particular it is now clear that a predicate may be observable of some kinds of entities and not of others. 'Spherical' is observable of baseballs (entrenched and directly and pragmatically observable), but not of protons; 'charged' is observable in at least some of these senses of pith balls but not of ions; and so on. No monadic predicate is observable of a theoretical entity; some predicates may be observable of some observable entities but not of others: for example, 'spherical' is not directly or pragmatically observable of the earth. The question whether there are absolutely theoretical *predicates* can now be seen to be independent of the question of theoretical entities; if there are none, this does not imply that there are no theoretical entities, not that predicates ascribed to them may not also be ascribed to observable entities.

How is a predicate ascribed to theoretical entities or to observable entities of which it is not itself observable? If it is a predicate that has already been ascribed directly to some observable entity, it may be inferred of another entity by analogical argument. For example, stones released near the surface of Jupiter will fall toward it because Jupiter is in other relevant respects like the earth. In the case of a theoretical entity, the analogical argument will have to involve relational predicates: high energy radiation arrives from a certain direction; it is inferred from other instances of observed radiation transmission between pairs of objects that there is a body at a certain point of space having a certain structure, temperature, gravitational field and so on.

But it is certain that some predicates have been introduced into science which do not appear in the relatively entrenched observation language. How are they predicated of objects? Consistently with the network model, there seem to be just two ways of introducing such newly minted predicates. First, they may be introduced as new observation predicates by assigning them to recognizable empirical situations where descriptions have not been required in prescientific language. Fairly clear examples are 'bacteria' when first observed in microscopes and 'sonic booms' first observed when aircraft 'broke the sound barrier'. Such introductions of novel terms will of course share the characteristic of all observation predicates of being dependent for their functions on observed associations or laws as well as direct empirical recognitions. In some cases it may be difficult to distinguish them from predicates introduced by *definition* in terms of previously familiar observation predicates. Fairly clear examples of this are 'molecule', defined as a small particle with certain physical and chemical properties such as mass, size, geometrical structure, and combinations and dissociations with other molecules, which are expressible in available predicates (most *names* of theoretical entities seem to be introduced this way); or 'entropy', defined quantitatively and operationally in terms of change of heat content divided by absolute temperature. In intermediate cases, such as 'virus', 'quasar' and 'Oedipus complex', it may be difficult to decide whether the function of these predicates is exhausted by logical equivalence with certain complex observation predicates or whether they can be said to have an independent function in some empirical situations where they are relatively directly observed. Such ambiguities are to be expected because, in the network model, laws which are strongly entrenched may sometimes be taken to be definitional, and laws introduced as definitions may later be regarded as being falsifiable empirical associations.

Notice that in this account the view of the function of predicates in theories that has been presupposed is explicitly nonformalist. The account is in fact closely akin to the view that all theories require to be interpreted in some relatively observable model, for in such a model their predicates are ascribed in observation statements. It has been assumed that when familiar predicates such as 'charge', 'mass' and 'position' are used of theoretical entities, these predicates are the 'same' as the typographically similar predicates used in observation statements. But it may be objected that when, say, elementary particles are described in terms of such predicates, the predicates are not used in their usual sense, for if they were, irrelevant models and analogies would be imported into the theoretical descriptions. It is important to be clear what this objection amounts to. If it is the assertion that a predicate such as 'charge' used of a theoretical entity has a sense related to that of 'charge' used of an observable entity only through the apparatus of formal deductive system plus correspondence rules, then the assertion is equivalent to a formal construal of theories, and it is not clear why the word 'charge' should be used at all. It would be less conducive to ambiguity to replace it with an uninterpreted sign related merely by the theoretical postulates and correspondence rules to observation predicates. If, however, the claim that it is used of theoretical entities in a different sense implies only that charged elementary particles are different kinds of entities from charged pith balls, this claim can easily be admitted and can be expressed by saying that they predicate co-occurs and is co-absent with different predicates in the two cases. The fact that use of the predicate has different lawlike implications in relatively theoretical contexts from those in observation contexts is better represented in the network model than in most other accounts of theories, for it has already been noticed that in this model the conditions of correct application of a predicate depend partly on the other predicates with which it is observed to occur. This seems sufficiently to capture what is in mind when it is asserted that 'charge' 'means' something different when applied to elementary particles and pith balls, or 'mass' when used in Newtonian and relativistic mechanics.

Since formalism has been rejected, we shall regard predicates such as those just described as retaining their identity (and hence their logical substitutivity) whether used of observable or theoretical entities, though they do not generally retain the same empirical situations of direct application. But the formalist account, even if rejected as it stands, does suggest another possibility for the introduction of new theoretical predicates, related to observation neither by assignment in recognizable empirical situations nor by explicit definition in terms of old predicates. Can the network model not incorporate new predicates whose relations with each other and with observation predicates are 'implicit', not in the sense intended by formalists, but rather as a new predicate might be coined in myth or in poetry, and understood in terms of its context, that is to say, of its asserted relations with both new and familiar predicates? This suggestion is perhaps nearer the intentions of some deductivists than is pure formalism from which it is insufficiently discriminated.[17]

It is not difficult to see how such a suggestion could be incorporated into the network model. Suppose instead of relating predicates by known laws, we *invent a myth* in which we describe entities in terms of some predicates already in the language, but in which we introduce other predicates in terms of some mythical situations and mythical laws. In other words we build up the network of predicates and laws partly imaginatively, but not in such a way as to contradict known laws, as in a good piece of science fiction.[18] It is, moreover, perfectly possible that such a system might turn out to have true and useful implications in the empirical domain of the original predicates, and in this way the mythical predicates and laws may come to have empirical reference and truth. This is not merely to repeat the formalist account of theoretical predicates as having meaning only in virtue of their place within a postulate system, because it is not necessary for such a formal system to have any interpretation, whereas here there is an interpretation, albeit an imaginary one. Neither are the predicates introduced here by any mysterious 'implicit definition' by a postulate system; they are introduced by the same two routes as are all other predicates, except that the laws and the empirical situations involved are imaginary.

Whether any such introduction of new predicates by mythmaking has ever occurred in science may be regarded as an open question. The opinion may be hazarded that no convincing examples have yet been identified. All theory construction, of course, involves an element of mythmaking, because it makes use of *familiar* predicates related in new ways by postulated laws not yet accepted as true. Bohr's atom, for example, was postulated to behave as no physical system had ever been known to behave; however, the entities involved were all described in terms of predicates already available in the language. There is, moreover, a reason why the mythical method of introducing new predicates is not likely to be very widespread in science. The reason is that use of known predicates which already contain some accepted lawlike implications allow inductive and analogical inference to further as yet unknown laws, which mythical predicates do not allow. There could be no prior inductive confidence in the implications of predicates and laws which were wholly mythical, as there can be in the implications of predicates at least some of whose laws are accepted. How important such inductive confidence is, however, is a controversial question which will be pursued later. But it is sufficient to notice that the network model does not demand that theories should be restricted to use of predicates already current in the language or observable in some domain of entities.

V. Theories

Under the guise of an examination of observational and theoretical predicates, I have in fact described a full-fledged account of theories, observation and the relation of the one to the other. This is of course to be expected, because the present account amounts to a denial that there is a fundamental distinction between theoretical and observation predicates and statements, and implies that the distinction commonly made is both obscure and misleading. It should not therefore be necessary to say much more about the place of theories in this account. I have so far tried to avoid the term 'theory', except when describing alternative views, and have talked instead about laws and lawlike implications. But a theory *is* just such a complex of laws and implications, some of which are well entrenched, others less so, and others again hardly more than suggestions with as yet little empirical backing. A given theory may in principle be formulated in various ways, and some such formulations will identify various of the laws with postulates; others with explicit definitions; others with theorems, correspondence rules or experimental laws. But the upshot of the whole argument is that these distinctions of function in a theory are relative not only to the particular formulation of the 'same' theory (as with various axiomatizations of mechanics or quantum theory) but also to the theory itself, so that what appears in one theory as an experimental law relating 'observables' may in another be a high-level theoretical postulate (think of the chameleon-like character of the law of inertia, or the conservation of energy). It is one of the more misleading results of the deductive account that the notion of 'levels', which has proper application to proofs in a formal postulate system in terms of the order of deducibility of theorems, has become transferred to an ordering in terms of more and less 'theory-laden' constituents of the theory. It should be clear from what has already been said that these two notions of 'level' are by no means co-extensive.

So much is merely the immediate application to theories of the general thesis here presented about descriptive predicates. But to drive the argument home, it will be as well to consider explicitly some of the problems which the theory–observation relation has traditionally been felt to raise and how they fare in the present account.

The circularity objection

An objection is sometimes expressed as follows: if the use of all observation predicates carries theoretical implications, how can they be used in descriptions which are claimed to be evidence for these same theories? At least it must be possible to find terms in which to express the evidence which are not laden with the theory for which they express the evidence.

This is at best a half-truth. If by 'theory-laden' is meant that the terms used in the observation report presuppose the *truth* of the very theory under test, then indeed this observation report cannot contribute evidence for this theory. If, for example, 'motion in a straight line with uniform speed' is *defined* (perhaps in a complex and disguised fashion) to be equivalent to 'motion under no forces', this definition implies the truth of the law of inertia, and an observation report to the effect that a body moving under no forces moves in a straight line with uniform speed does not constitute evidence for this law. The logic of this can be expressed as follows.

Definition: $P(x) \equiv df\ Q(x)$
Theory: $(x)[P(x) \supset Q(x)]$
Observation: $P(a) \& Q(a)$

Clearly neither theory nor observation report states anything empirical about the relation of P and Q.

Contrast this with the situation where the 'theory-loading' of $P(a)$ is interpreted to mean 'Application of P to an object a

implies acceptance of the truth of some laws into which *P* enters, and these laws are part of the theory under test', or, colloquially, 'The meaning of *P* presupposes the truth of some laws in the theory under test'. In the inertia example the judgment that *a* is a body moving in a straight line with uniform speed depends on the truth of laws relating measuring rods and clocks, the concept of 'rigid body', and ultimately on the physical truth of the postulates of Euclidean geometry, and possibly of classical optics. All these are part of the theory of Newtonian dynamics and are confirmed by the very same kinds of observation as those which partially justify the assertion $P(a)\&Q(a)$. The notion of an observation report in this account is by no means simple. It may include a great deal of other evidence besides the report that $P(a)\&Q(a)$, namely the truth of other implications of correct application of *P* to *a* and even the truth of universal laws of a high degree of abstractness. It is, of course, a standard objection to accounts such as the present, which have an element of 'coherence' in their criteria of truth, that nothing can be known to be true until everything is known. But although an adequate confirmation theory for our account would not be straightforward, it would be perfectly possible to develop one in which the correct applicability of predicates, even in observation reports, is strongly influenced by the truth of some laws into which they enter, and only vanishingly influenced by others. The notion of degrees of entrenchment relative to given theories would be essential to expressing the total evidence in such a confirmation theory.[19]

The reply to the circularity objection as it has been stated is, then, that although the 'meaning' of observation reports is 'theory-laden', the truth of particular theoretical statements depends on the coherence of the network of theory and its empirical input. The objection can be put in another way however: if the meaning of the terms in a given observation report is even partially determined by a theory for which this report is evidence, how can the same report be used to decide between two theories, as in the classic situation of a crucial experiment? For if this account is correct the same report cannot have the same meaning as evidence for two different theories.

This objection can be countered by remembering what it is for the 'meaning' of the observation report to be 'determined by the theory'. This entails that ascription of predicates in the observation report implies acceptance of various other laws relating predicates of the theory, and we have already agreed that there may be a hard core of such laws which are more significant for determining correct use than others. Now it is quite possible that two theories which differ very radically in most of their implications still contain some hard-core predicates and laws which they both share. Thus, Newtonian and Einsteinian dynamics differ radically in the laws into which the predicate 'inertial motion' enters, but they share such hard-core predicates as 'acceleration of falling bodies near the earth's surface', 'velocity of light transmitted from the sun to the earth', and so on, and they share some of the laws into which these predicates enter. It is this area of *intersection* of laws that must determine the application of predicates in the report of a crucial experiment. The situation of crucial test between theories is not correctly described in terms of 'withdrawal to a neutral observation language', because, as has already been argued, there is no such thing as an absolutely neutral or non-theory-laden language. It should rather be described as exploitation of the area of intersection of predicates and laws between the theories; this is, of course, entirely relative to the theories in question.

An example due originally to Feyerabend[20] may be developed to illustrate this last point. Anaximenes and Aristotle are devising a crucial experiment to decide between their respective theories of free fall. Anaximenes holds that the earth is disc-shaped and suspended in a non-isotropic universe in which there is a preferred direction of fall, namely the parallel lines perpendicular to and directed towards the surface of the disc on which Greece is situated. Aristotle, on the other hand, holds that the earth is a large sphere, much larger than the surface area of Greece, and that it is situated at the centre of a universe organized in a series of concentric shells, whose radii directed toward the centre determine the direction of fall at each point. Now clearly the word 'fall' as used by each of them is, in a sense, loaded with his own theory. For Anaximenes it refers to a preferred direction uniform throughout space; for Aristotle it refers to radii meeting at the centre of the earth. But equally clearly, while they both remain in Greece and converse on non-philosophical topics, they will use the word 'fall' without danger of mutual misunderstanding. For each of them the word will be correlated with the direction from his head to his feet when standing up, and with the direction from a calculable point of the heavens toward the Acropolis at Athens. Also, both of them will share most of these lawlike implications of 'fall' with ordinary Greek speakers, although the latter probably do not have any expectations about a preferred direction throughout universal space. This is not to say, of course, that the ordinary Greek speaker uses the word with *fewer* implications, for he may associate it with the passage from truth to falsehood, good to evil, heaven to hell —implications which the philosophers have abandoned.

Now suppose Anaximenes and Aristotle agree on a crucial experiment. They are blindfolded and carried on a Persian carpet to the other side of the earth. That it is the other side might be agreed upon by them, for example, in terms of star positions—this would be part of the intersection of their two theories. They now prepare to let go of a stone they have brought with them. Anaximenes accepts that this will be a test of his theory of fall and predicts, 'The stone will fall.' Aristotle accepts that this will be a test of his theory of fall and predicts, 'The stone will fall.' Their Persian pilot performs the experiment. Aristotle is delighted and cries, 'It falls! My theory is confirmed.' Anaximenes is crestfallen and mutters, 'It rises; my theory must be wrong.' Aristotle now notices that there is something strange about the way in which they have expressed their respective predictions and observation reports, and they embark upon an absorbing analysis of the nature of observation predicates and theory-ladenness.

The moral of this tale is simply that confirmation and refutation of competing theories does not depend on all observers using their language with the same 'meaning', nor upon the existence of any neutral language. In this case, of course, they could, if they thought of it, agree to make their predictions in terms of 'moves from head to foot', instead of 'falls', but *this* would have presupposed that men naturally stand with their feet on the ground at the Antipodes, and this is as much an uncertain empirical prediction as the original one. Even 'moves perpendicularly to the earth' presupposes that the Antipodes is not a series of steeply sloping enclosed caves and tunnels in which it is impossible to know whether the stars occasionally glimpsed are reflected in lakes or seen through gaps in thick clouds. In Anaximenes' universe, almost anything might happen. But we are not trying to show that in any particular example there are *no* intersections of theories, only that the intersection does not constitute an independent observation language, and that some predicates of the observation reports need not even lie in the intersection in order for testing and mutual understanding to be possible. The final analysis undertaken by Anaximenes and Aristotle will doubtless include the learning of each other's theories and corresponding predicates or the devising of a set of observation reports in the intersection of the two theories, or, more probably, the carrying out of both together.

As a corollary of this account of the intersections of theories, it should be noted that there is no *a priori* guarantee that two persons brought up in the same language community will use their words with the same meanings in all situations, even when each of them is conforming to standard logic within his own theory. If they discourse only about events which lie in the intersection of their theories, that they may have different theories will never be behaviourally detected. But such behavioural criteria for 'same meaning' may break down if their theories are in fact different and if they are faced with new situations falling outside the intersection. Misunderstanding and logical incoherence cannot be logically guarded against in the empirical use of language. The novelty of the present approach, however, lies not in that comparatively trivial remark, but in demonstrating that rational communication can take place in intersections, even when words are being used with 'different meanings', that is, with different implications in areas remote from the intersection.

The two-language account and correspondence rules

Many writers have seen the status of the so-called rules of interpretation, or correspondence rules, as the key to the proper understanding of the problem of theory and observation. The concept of correspondence rules presupposes the theory–observation distinction, which is bridged by the rules and therefore seems to have been bypassed in the present account. But there are cases where it seems so obvious that correspondence rules are both required and easily identifiable that it is necessary to give some attention to them, in case some features of the theory–observation relation have been overlooked.

These cases arise most persuasively where it seems to be possible to give two descriptions of a given situation, one in theoretical and one in observation terminology, and where the relation between these two descriptions is provided by the set of correspondence rules. Take, for example, the ordinary-language description of the table as hard, solid and blue, and the physicist's description of the same table in terms of atoms, forces, light waves and so on—or the familiar translation from talk of the pressure, volume and temperature of a gas to talk of the energy and momentum of random motions of molecules. It seems clear that in such examples there is a distinction between theoretical and observational descriptions and also that there are correspondence rules which determine the relations between them. How does the situation look in the network account?

It must be accepted at once that there is something more 'direct' about describing a table as hard, solid and blue than as a configuration of atoms exerting forces. 'Direct' is to be understood on our account in terms of the better entrenchment of the predicates 'hard', 'solid' and 'blue' and the laws which relate them, and in terms of the practical ease of learning and applying these predicates in the domain of tables, compared with the predicates of the physical description. This does not imply, however, that 'atom', 'force' and 'light wave' function in a distinct theoretical language nor that they require to be connected with observation predicates by extraneous and problematic correspondence rules. Consider as a specific example, usually regarded as a correspondence rule: '"This exerts strong repulsive forces" implies "This is hard".' Abbreviate this as 'Repulsion implies hardness', and call it *C*. What is the status of *C*? Various suggestions have been made, which I shall now examine.[21]

(a) It is an analytic definition.
This is an uninteresting possibility, and we shall assume it to be false, because 'repulsion' and 'hardness' are not synonymous in ordinary language. They are introduced in terms of different kinds of situation and generally enter into different sets of laws. Furthermore, in this domain of entities 'hardness' has the pragmatic characteristics of an observation predicate, and 'repulsion' of a theoretical predicate, and hence they cannot be synonymous here. Therefore, *C* is a synthetic statement.

How, then, do 'hard' and 'repulsion' function in the language? Consistently with our general account we should have to say something like this: meaning is given to 'hard' by a complex process of learning to associate the sound with certain experiences and also by accepting certain empirical correlations between occurrences reported as 'This is hard', 'This exerts pressure' (as of a spring or balloon), 'This is an area of strong repulsive force' (as of iron in the neighbourhood of a magnet), 'This is solid, impenetrable, undeformable . . .', 'This bounces, is elastic . . .'. Similarly, 'repulsion' is introduced in a set of instances including some of those just mentioned and also by means of Newton's second law and all its empirical instances. Granted that this is how we *understand* the terms of *C*, what kind of synthetic statement is it?

(b) It may be suggested that it is a theorem of the deductive system representing the physical theory.
This possibility has to be rejected by two-language philosophers because for them 'hard' does not occur in the language of the theory and, hence, cannot appear in any theorem of the theory. But for us the possibility is open, because both terms of *C* occur in the same language, and it is perfectly possible that having never touched tables, but knowing all that physics can tell us about the forces exerted by atoms and knowing also analogous situations in which repulsive forces are in fact correlated with the property of hardness (springs,

balloons, etc.), we may be able to deduce *C* as a theorem in this complex of laws.

(c) More simply, *C* may be not so much a deductive inference from a system of laws as an inductive or analogical inference from other accepted empirical correlations of repulsive force and hardness.
This is a possibility two-language philosophers are prone to overlook, because they are wedded to the notion that 'repulsion' is a theoretical term in the context of tables and therefore not a candidate for directly observed empirical correlations. But it does not follow that it is not comparatively observable in other domains—springs, magnets, and the like. Observation predicates are, as we have remarked, relative to a domain of entities.

(d) Unable to accept (a), (b) or (c), the two-language philosopher is almost forced to adopt yet another alternative in his account of correspondence rules, namely, that *C* is an independent empirical postulate,[22] which is added to the postulates of the theory to make possible the deduction of observable consequences from that theory.
There is no need to deny that this possibility may sometimes be exemplified. It should only be remarked that if all correspondence rules are logically bound to have this status, as a two-language philosopher seems forced to hold, some very strange and undesirable consequences follow. If there are no deductive, inductive or analogical reasons other than the physicist's fiat why particular theoretical terms should be correlated with particular observation terms, how is it possible for the 'floating' theory ever to be refuted? It would seem that we could always deal with an apparent refutation at the observation level by arbitrarily modifying the correspondence rules, for since on this view these rules are logically and empirically quite independent of the theory proper, they can always be modified without any disturbance to the theory itself. It may be replied that considerations of simplicity would prevent such arbitrary salvaging of a theory. But this objection can be put in a stronger form: it has very often been the case that a well-confirmed theory has enabled predictions to be made in the domain of observation, where

the deduction involved one or more *new* correspondence rules, relating theoretical to observation terms in a new way. If these correspondence rules were postulates introduced for no reason intrinsic to the theory, it is impossible to understand how such predictions could be made with confidence.

On the present account, then, it need not be denied that there is sometimes a useful distinction to be made between comparatively theoretical and comparatively observational descriptions, nor that there are some expressions with the special function of relating these descriptions. But this does not mean that the distinction is more than pragmatically convenient, nor that the correspondence rules form a logically distinct class of statements with unique status. Statements commonly regarded as correspondence rules may in different circumstances function as independent theoretical postulates, as theorems, as inductive inferences, as empirical laws, or even in uninteresting cases as analytic definitions. There is no one method of bridging a logical gap between theory and observation. There is no such logical gap.

Replaceability

Granted that there is a relative distinction between a set of less entrenched (relatively theoretical) predicates and better entrenched observation predicates, and that correspondence rules do not form a special class of statements relating these two kinds of predicates, there still remains the question: What is the relation between two descriptions of the same subject matter, one referring to theoretical entities and the other observation entities?

First of all, it follows from the present account that the two descriptions are not equivalent or freely interchangeable. To describe a table as a configuration of atoms exerting forces is to use predicates which enter into a system of laws having implications far beyond the domain of tables. The description of the table in ordinary language as hard and solid also has implications, which may not be fewer in number but are certainly different. One contrast between the two descriptions which should be noted is that the lawlike implications of the theoretical descriptions are much more explicit and unambiguous[23] than those of ordinary-language predicates like 'hard' and 'solid'. Because of this comparative imprecision it is possible to hold various views about the status of an observational description. It is sometimes argued that an observational description is straightforwardly *false*, because it carries implications contradicted by the theoretical description, which are probably derived from out-of-date science. Thus it is held that to say a table is hard and solid implies that it is a continuum of material substance with no 'holes' and that to touch it is to come into immediate contact with its substance. According to current physics these implications are false. Therefore, it is claimed, in all honesty we must in principle replace all our talk in observation predicates by talk in theoretical predicates in which we can tell the truth.

This view has a superficial attraction, but as it stands it has the very odd consequence that most the descriptions we ever give of the world are not only false but known to be false in known respects. While retaining the spirit of the replaceability thesis, this consequence can be avoided in two ways. First, we can make use of the notion of intersection of theories to remark that there will be a domain of discourse in which there is practical equivalence between some implications of observational description and some implications of theoretical description. In this domain the observation language user is telling the truth so long as he is not tempted to make inferences outside the domain. This is also the domain in which pragmatic observation reports provide the original evidence for the theory. Within this domain ordinary conversation can go on for a long time without its becoming apparent that an observation language user, an ordinary linguist and a theoretician are 'speaking different languages' in the sense of being committed to different implications outside the domain. It may even be the case that an ordinary linguist is not committed to *any* implications outside the domain which conflict with those of the theoretician. For example, and in spite of much argument to the contrary, it is not at all clear that the user of the ordinary English word 'solid' *is*, or ever was, committed to holding that a table is, in the mathematically infinitesimal, a continuum of substance. The question probably never occurred to him, either in the seventeenth century or the twentieth, unless he had been exposed to some physics. Secondly, the network account of predicates makes room for change in the function of predicates with changing knowledge of laws. In this case it may very well be that use of 'hard' and 'solid' in the observation language comes to have whatever implications are correct in the light of the laws of physics or else to have built-in limitations on their applicability, for example, to the domain of the very small.

These suggestions help to put in the proper light the thesis that it is in principle possible to replace the observation language by the theoretical, and even to teach the theoretical language initially to children without going through the medium of the observation language.[24] Such teaching may indeed be *in principle* possible, but consider what would happen if we assume that the children are being brought up in normal surroundings without special experiences devised by physicists. They will then learn the language in the intersection of physics and ordinary language; and though they may be taught to mouth such predicates as 'area of strong repulsive force' where other children are taught 'hard', they will give essentially the same descriptions in this intersection as the ordinary linguist, except that every observation predicate will be replaced by a string of theoretical predicates. Doubtless they will be better off when they come to learn physics, much of which they will have learned implicitly already; and if they were brought up from the start in a highly unnatural environment, say in a spaceship, even ordinary discourse might well be more conveniently handled in theoretical language. But all these possibilities do not seem to raise any special problems or paradoxes.

Explanation and reduction

It has been presupposed in the previous section that when an observational description and a theoretical description of the same situation are given, both have reference to the same entities, and that they can be said to contradict or to agree with one another. Furthermore, it has been suggested that there are circumstances in which the two descriptions may be equivalent, namely when both descriptions are restricted to a certain intersection of theoretical implications and when the implications of the observation predicates have been modified in the light of laws constituting the theory. Sometimes the objection is made to this account of the relation of theoretical and observational descriptions that, far from their being potentially equivalent descriptions of the same entities, the theory is intended to *explain* the observations; and explanation, it is held, must be given in terms which are different from what is to be explained. And, it is sometimes added, explanation must be a description of *causes* which are distinct from their observable effects.

It should first be noticed that this argument cannot be used in defence of the two-language view. That explanations are supposed to refer to entities different from those referred to in the explananda does not imply that these sets of entities have to be described in different languages. Explanation of an accident, a good crop or an economic crisis will generally be given in the same language as that of the explananda.

It does seem, however, that when we give the theoretical description of a table as a configuration of atoms exerting repulsive forces, we are saying something which *explains* the fact that the table is hard and states the *causes* of that hardness. How then can this description be in any sense *equivalent* to the observational description of the table as hard? It does of course follow from the present account that they are not equivalent in the sense of being *synonymous*. That much is implied by the different function of the theoretical and observation predicates. Rather, the descriptions are equivalent in the sense of having the same reference, as the morning star is an alternative description of the evening star and also of the planet Venus. It is possible for a redescription in this sense to be explanatory, for the redescription of the table in theoretical terms serves to place the table in the context of all the laws and implications of the theoretical system. It is not its reference to the *table* that makes it explanatory of the observation statements, which also have reference to the table. It is rather explanatory because it says of the table that in being 'hard' ('exerting repulsive force') it is *like* other objects which are known to exert repulsive force and to feel hard as the table feels hard, and that the table is, therefore, an instance of general laws relating dynamical properties with sensations. And in regard to the causal aspects of explanation, notice that the repulsive forces are not properly said to be the causes of the table having the property 'hard', for they *are* the property 'hard'; but rather repulsive forces are causes of the table *feeling* hard, where 'hard' is not a description of the table but of a sensation. Thus the cause is not the same as the effect, because the referents of the two descriptions are different; and the explanans is not the same as the explanandum, because although their referents are the same, the theoretical description explains by relating the explanandum to other like entities and to a system of laws, just in virtue of its use of relatively theoretical predicates. There is some truth in the orthodox deductive account of explanation as deducibility in a theoretical system, but there is also truth in the contention that explanation involves stating as well what the explanandum *really* is and, hence, relating it to other systems which are then seen to be essentially similar to it. Initial misdescription of the function of descriptive predicates precludes the deductive account from doing justice to these latter aspects of explanation, whereas in the present account they are already implied in the fact that redescription in theoretical predicates carries with it lawlike relations between the explanandum and other essentially similar systems.

VI. Conclusion

In this chapter I have outlined a network model of theoretical science and argued that it represents the structure of science better than the traditional deductivist account, with its accompanying distinction between the theoretical and the observational.

First, I investigated some consequences of treating the theoretical and the observational aspects of science as equally problematic from the point of view of truth conditions and meaning. I described the application of observation in empirical situations as a classificatory process, in which unverbalized empirical information is lost. Consequently, reclassification may in principle take place in any part of the observational domain, depending on what internal constraints are imposed by the theoretical network relating the observations. At any given stage of science there are *relatively* entrenched observation statements, but any of these may later be rejected to maintain the economy and coherence of the total system.

This view has some similarity with other non-deductivist accounts in which observations are held to be 'theory-laden', but two familiar objections to views of this kind can be answered more directly in the network account. First, it is not a conventionalist account in the sense that any theory can be imposed upon any facts regardless of coherence conditions. Secondly, there is no vicious circularity of truth and meaning, for at any given time *some* observation statements result from correctly applying observation terms to empirical situations according to learned precedents and independently of theories, although the relation of observation and theory is a self-correcting process in which it is not possible to know at the time which of the set of observation statements are to be retained as correct in this sense, because subsequent observations may result in rejection of some of them.

Turning to the relatively theoretical aspects of science, I have argued that a distinction should be made between theoretical *entities* and theoretical *predicates*. I have suggested that if by theoretical predicates is meant those which are never applied in observational situations to any objects, and if the open-ended character of even observation predicates is kept in mind, there are no occasions on which theoretical predicates are used in science, although of course there are many theoretical entities to which predicates observable in other situations are applied. It follows that there is no distinction in kind between a theoretical and an observation language. Finally, I have returned to those aspects of scientific theories which are analysed in the deductive view in terms of the alleged theory–observation distinction and shown how they can be reinterpreted in the network model. Correspondence rules become empirical relations between relatively theoretical and relatively observational parts of the network; replaceability of observational descriptions by theoretical descriptions becomes redescription in more general terms in which the 'deep' theoretical similarities between observationally diverse systems are revealed; and theoretical explanation is understood similarly as redescription and not as causal relationship between distinct theoretical and observable domains of entities mysteriously inhabiting the same space–time region. Eddington's two tables are one table.

Notes

1 A. J. Ayer, *Language, Truth and Logic*, 2nd edn, London, 1946, p. 11; R. B. Braithwaite, *Scientific Explanation*, New York, 1953, p. 8; R. Carnap, 'The methodological character of theoretical concepts', in *Minnesota Studies in the Philosophy of Science*, vol. i, ed. H. Feigl and M. Scriven, Minneapolis, 1956, p. 38; C. G. Hempel, 'The theoretician's dilemma' in *Minnesota Studies in the Philosophy of Science*, vol. ii, ed. H. Feigl, M. Scriven and G. Maxwell, Minneapolis, 1958, p. 41.

2 E. Nagel, *The Structure of Science*, New York and London, 1961, ch. 5.

3 It would be impossible to give examples from sciences other than physics: 'adaptation', 'function', 'intention', 'behaviour', 'unconscious mind'; but the question whether these are theoretical terms in the sense here distinguished from observation terms is controversial, and so is the question whether, if they are, they are eliminable from their respective sciences. These questions would take us too far afield.

4 D. Davidson, 'Theories of meaning and learnable language' in *Logic, Methodology and Philosophy of Science*, ed. Y. Bar-Hillel, Amsterdam, 1965, p. 386.

5 N. Chomsky, 'Quine's empirical assumptions', *Synthese*, vol. xix, 1968, 53; see also Quine's reply to Chomsky, *ibid.*, 274.

6 K. R. Popper, *The Logic of Scientific Discovery*, London, 1959, appendix x, p. 422.

7 See for example, L. Wittgenstein, *Philosophical Investigations*, London, 1953, section 258 ff; A. J. Ayer, *The Concept of a Person*, London, 1963, pp. 39 ff; Popper, *op. cit.*, pp. 44–5.

8 Cf. P. K. Feyerabend, 'An attempt at a realistic interpretation of experience', *Proc. Arist. Soc.* vol. lviii, 1957–8, 143, 160.

9 N. R. Campbell, *Foundations of Science*, Cambridge, 1920, p. 29.

10 That this way of putting it is a gross distortion of Einstein's actual thought processes is irrelevant here.

11 The many-dimensional character of the theory–observation distinction has been discussed by P. Achinstein in 'The problem of theoretical terms', *Am. Phil. Quart.*, vol. ii, 1965, 193 and *Concepts of Science*, Baltimore, 1968, chs 5 and 6.

12 P. Duhem, *The Aim and Structure of Physical Theory*, (first published as *La Théorie Physique*, 1906; English translation, Cambridge, Mass., 1953, p. 133.

13 W. v. O. Quine, *From a Logical Point of View*, Cambridge, Mass., 1953, p. 43.

14 W. v. O. Quine, *Word and Object*, Cambridge, Mass., 1953, pp. 42, 43.

15 *Ibid.*, p. 11. For an early and devastating investigation of the notion of 'implicit definition' in a formal system, see G. Frege, 'On the foundations of geometry' (first published 1903), trans. M. E. Szabo, *Philosophical Review*, vol. lxix, 1960, 3, and in specific relation to the deductive account of theories, see C. G. Hempel, 'Fundamentals of concept formation in empirical science', *International Encyclopaedia of Unified Science*, vol. ii, no. 7, Chicago, 1952, p. 81.

16 Hempel, 'Fundamentals of concept formation', p. 36.

17 It certainly represents what Quine seems to have *understood* some deductive accounts to be, cf. above, pp. 86–7.

18 We build up the network in somewhat the same way that M. Black in *Models and Metaphors*, Ithaca, 1962, p. 43, suggests a poet builds up a web of imagined associations within the poem itself in order to make new metaphors intelligible. He might, indeed, in this way actually coin and give currency to wholly new words.

19 That no simple formal examples can be given of the present account is perhaps one reason why it has not long ago superseded the deductive account. I have made some preliminary suggestions towards a confirmation theory for the network account in my chapter on 'Positivism and the logic of scientific theories' in *The Legacy of Logical Positivism for the Philosophy of Science*, ed. P. Achinstein and S. Barker, Baltimore, 1969, p. 85, and in 'A self-correcting observation language' in *Logic, Methodology and Philosophy of Science*, ed. B. van Rootselaar and J. F. Stahl, Amsterdam, 1968, p. 297. These suggestions are developed below, chs 5–9.

20 P. K. Feyerabend, 'Explanation, reduction and empiricism', *Minnesota Studies in the Philosophy of Science*, vol. iii, ed. H. Feigl and G. Maxwell, Minneapolis, 1962, 85.

21 I owe several of these suggestions to a private communication from Paul E. Meehl; see also Nagel, *Structure of Science*, pp. 354 ff.

22 Cf. Carnap's 'meaning postulates' in *Philosophical Foundations of Physics*, New York and London, 1966, ch. 27.

23 This is not to say, as formalists are prone to do, that theoretical terms are completely unambiguous, precise or exact, like the terms of a formal system. If they were, this whole account of the functioning of predicates would be mistaken. The question of precision deserves more extended discussion. It has been investigated by Stephan Körner, *Experience and Theory*, London, 1966; D. H. Mellor, 'Experimental error and deducibility', *Philosophy of Science*, vol. xxxii, 1965, 105 and *idem.*, 'Inexactness and explanation', *ibid.*, vol. xxxiii, 1966. 345.

24 Cf. W. Sellars, 'The language of theories' in *Current Issues in the Philosophy of Science*, ed. H. Feigl and G. Maxwell, New York, 1961, p. 57.

4 *The Explanatory Function of Metaphor*

The thesis of this paper is that the deductive model of scientific explanation should be modified and supplemented by a view of theoretical explanation as metaphoric redescription of the domain of the explanandum. This raises two large preliminary questions: first, whether the deductive model requires modification, and second, what is the view of metaphor presupposed by the suggested alternative. I shall not discuss the first question explicitly. Much recent literature in the philosophy of science has answered it affirmatively,[1] and I shall refer briefly at the end to some difficulties tending to show that a new model of explanation is required, and suggest how the conception of theories as metaphors meets these difficulties.

The second question, about the view of metaphor presupposed, requires more extensive discussion. The view I shall present is essentially due to Max Black, who has developed in two papers, entitled respectively 'Metaphor' and 'Models and Archetypes',[2] both a new theory of metaphor, and a parallelism between the use of literary metaphor and the use of models in theoretical science. I shall start with an exposition of Black's *interaction view* of metaphors and models, taking account of modifications suggested by some of the subsequent literature on metaphor.[3] It is still unfortunately necessary to argue that metaphor is more than a decorative literary device, and that it has cognitive implications whose nature is a proper subject of philosophic discussion. But space forces me to mention these arguments as footnotes to Black's view, rather than as an explicit defence *ab initio* of the philosophic importance of metaphor.

The interaction view of metaphor

1. We start with two systems, situations, or referents, which will be called respectively the primary and secondary systems. Each is described in literal language. A metaphoric use of language in describing the primary system consists of

transferring to it a word or words normally used in connection with the secondary system: for example, 'Man is a wolf', 'Hell is a lake of ice'. In a scientific theory the primary system is the domain of the explanandum, describable in observation language; and the secondary is the system, described either in observation language or the language of a familiar theory, from which the model is taken: for example, 'sound (primary system) is propagated by wave motion (taken from a secondary system)'; 'gases are collections of randomly moving massive particles'.

Three terminological remarks should be inserted here. First, 'primary' and 'secondary system', and 'domain of the explanandum' will be used throughout to denote the referents or putative referents of descriptive statements; and 'metaphor', 'model', 'theory', 'explanans' and 'explanandum' will be used to denote linguistic entities. Second, use of the terms 'metaphoric' and 'literal', 'theory' and 'observation', need not be taken at this stage to imply a pair of irreducible dichotomies. All that is intended is that the 'literal' and 'observation' languages are assumed initially to be well understood and unproblematic, whereas the 'metaphoric' and 'theoretical' are in need of analysis. The third remark is that to assume initially that the two systems are 'described' in literal or observation language does not imply that they are exhaustively or accurately described or even that they could in principle be so in terms of these languages.

2. We assume that the primary and secondary systems each carry a set of associated ideas and beliefs that come to mind when the system is referred to. These are not private to individual language-users, but are largely common to a given language community and are presupposed by speakers who intend to be understood in that community. In literary contexts the associations may be loosely knit and variable, as in the wolf-like characteristics which come to mind when the metaphor 'Man is a wolf' is used; in scientific contexts the primary and secondary systems may both be highly organized by networks of natural laws.

A remark must be added here about the use of the word 'meaning'. Writers on metaphor appear to intend it as an inclusive term for reference, use, and the relevant set of associated ideas. It is, indeed, part of their thesis that it has to be understood thus widely. To understand the meaning of a descriptive expression is not only to be able to recognize its referent, or even to use the words in the expression correctly, but also to call to mind the ideas, both linguistic and empirical, which are commonly held to be associated with the referent in the given language community. Thus a shift of meaning may result from a change in the set of associated ideas, as well as in change of reference or use.

3. For a conjunction of terms drawn from the primary and secondary systems to constitute a metaphor it is necessary that there should be patent falsehood or even absurdity in taking the conjunction literally. Man is not, literally, a wolf; gases are not in the usual sense collections of massive particles. In consequence some writers have denied that the referent of the metaphoric expression can be identified with the primary system without falling into absurdity or contradiction. I shall return to this in the next section.

4. There is initially some principle of assimilation between primary and secondary systems, variously described in the literature as 'analogy', 'intimations of similarity', 'a programme for exploration', 'a framework through which the primary is seen.' Here we have to guard against two opposite interpretations, both of which are inadequate for the general understanding of metaphors and scientific models. On the one hand, to describe this ground of assimilation as a *programme* for exploration, or a *framework* though which the primary is seen, is to suggest that the secondary system can be imposed *a priori* upon the primary, as if *any* secondary can be the source of metaphors or models for *any* primary, provided the right metaphor-creating operations are subsequently carried out. Black does indeed suggest that in some cases 'it would be more illuminating . . . to say that the metaphor creates the similarity than to say it formulates some similarity antecedently existing' (p. 37), and he also points out that some poetry creates new metaphors precisely by itself developing the system of associations in terms of which

'absurd' conjunctions of words are to be metaphorically understood. There is however an important distinction to be brought out between such a use of metaphor and scientific models, for, whatever may be the case for poetic use, the suggestion that *any* scientific model can be imposed *a priori* on *any* explanandum and function fruitfully in its explanation must be resisted. Such a view would imply that theoretical models are irrefutable. That this is not the case is sufficiently illustrated by the history of the concept of a heat fluid, or the classical wave theory of light. Such examples also indicate that no model even gets off the ground unless some antecedent similarity or analogy is discerned between it and the explanandum.

But here there is a danger of falling into what Black calls the *comparison* view of metaphor. According to this view the metaphor can be replaced without remainder by an explicit, literal statement of the similarities between primary and secondary systems, in other words, by a simile. Thus, the metaphor 'Man is a wolf' would be equivalent to 'Man is like a wolf in that. . . .', where follows a list of comparable characteristics; or, in the case of theoretical models, the language derived from the secondary system would be wholly replaced by an explicit statement of the analogy between secondary and primary systems, after which further reference to the secondary system would be dispensable. Any interesting examples of the model-using in science will show, however, that the situation cannot be described in this way. For one thing, as long as the model is under active consideration as an ingredient in an explanation, we do not know how far the comparison extends—it is precisely in its extension that the fruitfulness of the model may lie. And a more fundamental objection to the comparison view emerges in considering the next point.

5. The metaphor works by transferring the associated ideas and implications of the secondary to the primary system. These select, emphasize, or suppress features of the primary; new slants on the primary are illuminated; the primary is 'seen through' the frame of the secondary. In accordance with the doctrine that even literal expressions are understood

partly in terms of the set of associated ideas carried by the system they describe, it follows that the associated ideas of the primary are changed to some extent by the use of the metaphor, and that therefore even its original literal description is shifted in meaning. The same applies to the secondary system, for its associations come to be affected by assimilation to the primary; the two systems are seen as more like each other; they seem to interact and adapt to one another, even to the point of invalidating their original literal descriptions if these are understood in the new, post-metaphoric sense. Men are seen to be more like wolves after the wolf-metaphor is used, and wolves seem to be more human. Nature becomes more like a machine in the mechanical philosophy, and actual, concrete machines themselves are seen as if stripped down to their essential qualities of mass in motion.

This point is the kernel of the interaction view, and is Black's major contribution to the analysis of metaphor. It is incompatible with the comparison view, which assumes that the literal descriptions of both systems are and remain independent of the use of the metaphor, and that the metaphor is reducible to them. The consequences of the interaction view for theoretical models are also incompatible with assumptions generally made in the deductive account of explanation, namely that descriptions and descriptive laws in the domain of the explanandum remain empirically acceptable and invariant in meaning to all changes of explanatory theory. I shall return to this point.

6. It should be added as a final point in this preliminary analysis that a metaphoric expression used for the first time, or used to someone who hears it for the first time, is intended to be *understood*. Indeed it may be said that a metaphor is not metaphor but nonsense if it communicates nothing, and that a genuine metaphor is also capable of communicating something other than was intended and hence of being *mis*understood. If I say (taking two words more or less at random from a dictionary page) 'A truck is a trumpet' it is unlikely that I shall communicate anything; if I say 'He is a shadow on the weary land', you may understand me to mean (roughly) 'He is a wet blanket, a gloom, a menace', whereas I actually

meant (again roughly) 'He is a shade from the heat, a comfort, a protection'.

Acceptance of the view that metaphors are meant to be intelligible implies rejection of all views which make metaphor a wholly non-cognitive, subjective, emotive, or stylistic use of language. There are exactly parallel views of scientific models which have been held by many contemporary philosophers of science, namely that models are purely subjective, psychological, and adopted by individuals for private heuristic purposes. But this is wholly to misdescribe their function in science. Models, like metaphors, are intended to communicate. If some theorist develops a theory in terms of a model, he does not regard it as a private language, but presents it as an ingredient of his theory. Neither can he, nor need he, make literally explicit all the associations of the model he is exploiting; other workers in the field 'catch on' to its intended implications, indeed they sometimes find the theory unsatisfactory just because some implications which the model's originator did not investigate, or even think of, turn out to be empirically false. None of this would be possible unless use of the model were intersubjective, part of the commonly understood theoretical language of science, not a private language of the individual theorist.

An important general consequence of the interaction view is that it is not possible to make a distinction between literal and metaphoric descriptions merely by asserting that literal use consists in the following of linguistic rules. Intelligible metaphor also implies the existence of rules of metaphoric use, and since in the interaction view literal meanings are shifted by their association with metaphors, it follows that the rules of literal usage and of metaphor, though they are not identical, are nevertheless not independent. It is not sufficiently clear in Black's paper that the interaction view commits one to the abandonment of a two-tiered account of langauge in which some usages are irreducibly literal and others metaphoric. The interaction view sees language as dynamic: an expression initially metaphoric may become literal (a 'dead' metaphor), and what is at one time literal may become metaphoric (for example the Homeric 'he breathed forth his life', originally literal, is now a metaphor for death). What is important is not to try to draw a line between the metaphoric and the literal, but rather to trace out the various mechanisms of meaning-shift and their interactions. The interaction view cannot consistently be made to rest on an initial set of absolutely literal descriptions, but rather on a relative distinction of literal and metaphoric in particular contexts. I cannot undertake the task of elucidating these conceptions here (an interesting attempt to do so has been made by K. I. B. S. Needham[4]), but I shall later point out a parallel between this general linguistic situation and the relative distinctions and mutual interactions of theory and observation in science.

The problem of metaphoric reference

One of the main problems for the interaction view in its application to theoretical explanation is the question what is the *referent* of a model or metaphor. At first sight the referent seems to be the primary system, which we choose to describe in metaphoric rather than literal terms. This, I believe, is in the end the right answer, but the process of metaphoric description is such as to cast doubt on any simple identification of the metaphor's reference with the primary system. It is claimed in the interaction view that a metaphor causes us to 'see' the primary system differently, and causes the meanings of terms originally literal in the primary system to shift towards the metaphor. Thus 'Man is a wolf' makes man seem more vulpine, 'Hell is a lake of ice' makes hell seem icy rather than hot, and a wave theory of sound makes sound seem more vibrant. But how can initial similarities between the objective systems justify such changes in the meanings of words and even, apparently, in the things themselves? Man does not in fact change because someone uses the wolf-metaphor. How then can we be justified in identifying what we see through the framework of the metaphor with the primary system itself? It seems that we cannot be entitled to say men *are* wolves, sound *is* wave motion, in any identificatory sense of the copula.

Some recent writers on metaphor[5] have made it the main

burden of their argument to deny that any such identification is possible. They argue that if we allow it we are falling into the absurdity of conjoining two literally incompatible systems, and the resulting expression is not metaphoric but meaningless. By thus taking a metaphor literally we turn it into a *myth*. An initial misunderstanding may be removed at once by remarking that 'identification' cannot mean in this context identification of the referent of the metaphoric expression, taken in its *literal* sense, with the primary system. But if the foregoing analysis of metaphor is accepted, then it follows that metaphoric use is use in a different from the literal sense, and furthermore it is use in a sense not replaceable by any literal expression. There remains the question what it is to identify the referent of the metaphoric expression or model with the primary system. As a preliminary to answering this question it is important to point out that there are two ways, which are often confused in accounts of the 'meaning of theoretical concepts', in which such identification may fail. It may fail because it is in principle meaningless to make any such identification, or it may fail because in a particular case the identification happens to be *false*. Instances of false identification, e.g. 'heat is a fluid' or 'the substance emitted by a burning object is phlogiston', provide no arguments to show that other such identifications may not be both meaningful and true.

Two sorts of argument have been brought against the view that metaphoric expressions and models can refer to and truly describe the primary system. The first depends on an assimilation of poetic and scientific metaphor, and points out that it is characteristic of good poetic metaphor that the images introduced are initially striking and unexpected, if not shocking; that they are meant to be entertained and savoured for the moment and not analyzed in pedantic detail nor stretched to radically new situations; and that they may immediately give place to other metaphors referring to the same subject matter which are formally contradictory, and in which the contradictions are an essential part of the total metaphoric impact. Any attempt to separate these literal contradictions from the nexus of interactions is destructive of the metaphor, particularly on the interaction view. In the light of these characteristics there is indeed a difficult problem about the correct analysis of the notion of metaphoric 'truth' in poetic contexts. Scientific models, however, are fortunately not so intractable. They do not share any of the characteristics listed above which make poetic metaphors peculiarly subject to formal contradictoriness. They may initially be unexpected, but it is not their chief aim to shock; they are meant to be exploited energetically and often in extreme quantitative detail and in quite novel observational domains; they are meant to be internally tightly knit by logical and causal interrelations; and if two models of the same primary system are found to be mutually inconsistent, this is not taken (*pace* the complementarity interpretation of quantum physics) to enhance their effectiveness, but rather as a challenge to reconcile them by mutual modification or to refute one of them. Thus their truth criteria, although not rigorously formalizable, are at least much clearer than in the case of poetic metaphor. We can perhaps signalize the difference by speaking in the case of scientific models of the (perhaps unattainable) aim to find a 'perfect metaphor', whose referent is the domain of the explanandum, whereas literary metaphors, however adequate and successful in their own terms, are from the point of view of potential logical consistency and extendability often (not always) intentionally imperfect.

Secondly, if the interaction view of scientific metaphor or model is combined with the claim that the referent of the metaphor is the primary system (i.e. the metaphor is true of the primary system), then it follows that the thesis of meaning-invariance of the literal observation-descriptions of the primary system is false. For, the interaction view implies that the meaning of the original literal language of the primary system is changed by adoption of the metaphor. Hence those who wish to adhere to meaning-invariance in the deductive account of explanation will be forced to reject either the interaction view or the realistic view that a scientific model is putatively true of its primary system. Generally they reject both. But abandonment of meaning-invariance, as in many recent criticisms of the deductive model of explanation, leaves room for adoption of both the interaction view, and realism, as I shall now try to spell out in more detail.

Explanation as metaphoric redescription

The initial contention of this paper was that the deductive model of explanation should be *modified* and *supplemented* by a view of theoretical explanation as metaphoric redescription of the domain of the explanandum. First, the association of the ideas of 'metaphor' and of 'explanation' requires more examination. It is certainly not the case that all explanations are metaphoric. To take only two examples, explanation by covering-law, where an instance of an *A* which is *B* is explained by reference to the law 'All *A*'s are *B*'s', is not metaphoric, neither is the explanation of the working of a mechanical gadget by reference to an actual mechanism of cogs, pulleys, and levers. These, however, are not examples of *theoretical* explanation, for it has been taken for granted that the essence of a theoretical explanation is the introduction into the explanans of a new vocabulary or even of a new language. But introduction of a metaphoric terminology is not in itself explanatory, for in literary metaphor in general there is no hint that what is metaphorically described is also thereby explained. The connection between metaphor and explanation is therefore neither that of necessary nor sufficient condition. Metaphor becomes explanatory only when it satisfies certain further conditions.

The orthodox deductive criteria for a scientific explanans[6] require that the explanandum be deducible from it, that it contain at least one general law not redundant to the deduction, that it be not empirically falsified up to date, and that it be predictive. We cannot simply graft these requirements on to the account of theories as metaphors without investigating the consequences of the interaction view of metaphor for the notions of 'deducibility', 'explanandum', and 'falsification' in the orthodox account. In any case, as has been mentioned already, the requirement of deducibility in particular has been subjected to damaging attack, quite apart from any metaphoric interpretation of theories. There are two chief grounds for this attack, both of which can be turned into arguments favourable to the metaphoric view.

In the first place it is pointed out that there is seldom in fact a deductive relation strictly speaking between scientific explanans and explanandum, but only relations of approximate fit. Furthermore, what counts as sufficiently approximate fit cannot be decided deductively, but is a complicated function of coherence with the rest of a theoretical system, general empirical acceptability throughout the domain of the explanandum, and many other factors. I do not propose to try to spell out these relationships in further detail here, but merely to make two points which are relevant to my immediate concern. First, the attack on deducibility drawn from the occurrence of approximations does not imply that there are *no* deductive relations between explanans and explanandum. The situation is rather this. Given a descriptive statement *D* in the domain of the explanandum, it is usually the case that the statement *E* of an acceptable explanans does not entail *D*, but rather *D′*, where *D′* is a statement in the domain of the explanandum only 'approximately equivalent' to *D*. For *E* to be acceptable it is necessary both that there be a deductive relation between *E* and *D′*, and that *D′* should come to be recognized as a *more acceptable* description in the domain of the explanandum than *D*. The reasons why it might be more acceptable—repetition of the experiments with greater accuracy, greater coherence with other acceptable laws, recognition of disturbing factors in arriving at *D* in the first place, metaphoric shifts in the meanings of terms in *D* consequent upon the introduction of the new terminology of *E*, and so on—need not concern us here. What is relevant is that the non-deducibility of *D* from *E* does not imply total abandonment of the deductive model unless *D* is regarded as an invariant description of the explanandum, automatically rendering *D′* empirically false. That *D* cannot be so regarded has been amply demonstrated in the literature. The second point of contact between these considerations and the view of theories as metaphors is now obvious. That explanation may modify and correct the explanandum is already built into the relation between metaphors and the primary system in the interaction view. Metaphors, if they are good ones, and *ipso facto* their deductive consequences, do have the primary system as their referents, for they may be seen as correcting and replacing the original literal descriptions of the same system, so that the literal descriptions are discarded as inadequate or even false. The parallel with the deductive relations of explanans and explananda is clear: the metaphoric view does not abandon deduction, but it focuses attention rather on the interaction between metaphor and primary system, and on the criteria of acceptability of metaphoric descriptions of the primary system, and hence not so much upon the deductive relations which appear in this account as comparatively uninteresting piece of logical machinery.

The second attack upon the orthodox deductive account gives even stronger and more immediate grounds for the introduction of the metaphoric view. It is objected that there are no deductive relations between theoretical explanans and explanandum because of the intervention of correspondence rules. If the deductive account is developed, as it usually is, in terms either of an uninterpreted calculus and an observation language, or of two distinct languages, the theoretical and the observational, it follows that the correspondence rules linking terms in these languages cannot be derived deductively from the explanans alone. Well-known problems then arise about the status of the correspondence rules and about the meaning of the predicates of the theoretical language. In the metaphoric view, however, these problems are evaded, because here there are no correspondence rules, and this view is primarily designed to give its own account of the meaning of the language of the explanans. There is *one* language, the observation language, which like all natural languages is continually being extended by metaphoric uses, and hence yields the terminology of the explanans. There is no problem about connecting explanans and explanandum other than the general problem of understanding how metaphors are introduced and applied and exploited in their primary systems. Admittedly, we are as yet far from understanding this process, but to see the problem of the 'meaning of theoretical concepts' as a special case is one step in its solution.

Finally, a word about the requirement that an explanation be predictive. It has been much debated within the orthodox deductive view whether this is a necessary and sufficient condition for explanation, and it is not appropriate here to enter into that debate. But any account of explanation would be inadequate which did not recognize that, in general, an explanation is required to be predictive, or, what is closely connected with this, to be falsifiable. Elsewhere[7] I have pointed out that, in terms of the deductive view, the requirement of predictivity may mean one of three things:

(i) That general laws already present in the explanans have as yet unobserved instances. This is a trivial fulfilment of the requirement, and would not, I think, generally be regarded as sufficient.

(ii) That further general laws can be derived from the explanans, *without* adding further items to the set of correspondence rules. That is to say, predictions remain within the domain of the set of predicates already present in the explanandum. This is a weak sense of predictivity which covers what would normally be called *applications* rather than extensions of a theory (for example, calculation of the orbit of a satellite from the theory of gravitation, but not extension of the theory to predict the bending of light rays).

(iii) There is also a strong sense of prediction in which new observation predicates are involved, and hence, in terms of the deductive view, additions are required to the set of correspondence rules.

I have argued[8] that there is no rational method of adding to the correspondence rules on the pure deductive view, and hence that cases of strong prediction cannot be rationally accounted for on that view. In the metaphoric view, on the other hand, since the domain of the explanandum is redescribed in terminology transferred from the secondary system, it is to be expected that the original observation language will both be shifted in meaning and extended in vocabulary, and hence that predictions in the strong sense will become possible. They may not of course turn out to be *true*, but that is an occupational hazard of any explanation or prediction. They will however be rational, because rationality consists just in the continuous adaptation of our language to our continually expanding world, and metaphor is one of the chief means by which this is accomplished.

Notes

1 See, for example, P. K. Feyerabend, 'An attempt at a realistic interpretation of experience', *Proc. Arist. Soc.*, vol. lviii, 1957, 143; *idem*, 'Explanation, reduction and empiricism' in *Minnesota Studies in the Philosophy of Science*, vol. iii, ed. H. Feigl and G. Maxwell, Minneapolis, 1962; T. S. Kuhn, *The Structure of Scientific Revolutions*, Chicago, 1962; W. Sellars, 'The language of theories' in *Current Issues in the Philosophy of Science*, ed. H. Feigl and G. Maxwell, New York, 1961.

2 Max Black, *Models and Metaphors*, Ithaca, 1962.

3 See M. C. Beardsley, *Aesthetics*, New York, 1958; D. Berggren, 'The use and abuse of metaphor', *Rev. Met.*, vol. xvi, 1962, 237 and 450; Mary A. McCloskey, 'Metaphors', *Mind*, vol. lxxiii, 1964, 215; D. Schon, *The Displacement of Concepts*, London, 1963; C. Turbayne, *The Myth of Metaphor*, New Haven, 1962.

4 K. I. B. S. Needham, 'Synonymy and semantic classification', unpublished Ph.D. thesis, Cambridge, 1964.

5 Bergren, McCloskey and Turbayne, in the works cited above.

6 For example C. G. Hempel and P. Oppenheim, 'The logic of explanation', reprinted in *Readings in the Philosophy of Science*, ed. H. Feigl and M. Brodbeck, New York, 1953, 319.

7 M. Hesse, *Models and Analogies in Science*, London, 1963; see also *idem*, 'A new look at scientific explanation', *Rev. Met.*, vol. xvii, 1963, 98.

8 M. Hesse, *ibid*, and 'Theories, dictionaries and observation', *British Journal of the Philosophy of Science*, vol. ix, 1958, 12 and 128.

5 *Models of Theory-change*

I

Science is essentially a learning device, and all models of the structure of science describe what are essentially mechanisms of learning. Almost without exception these models presuppose that the subject matter of learning is the empirical world, and that the world interacts with the learning device, whether that device is conceived as the individual scientist, the whole body of scientists, or the institution of science represented by its learned societies, journals, and textbooks. Again, almost without exception, models of science presuppose that the learning process returns to the empirical world, which provides checks and reinforcements, and is the subject of prediction and control. In terms of the mechanisms of learning intervening between empirical input and predictive output, almost all models of science, and indeed all epistemologies, can be located.

I propose to describe a very general type of learning machine in which this epistemological exercise can be carried out.[1] Consider a learning machine that goes through the following stages of operation:

1. *Empirical input* from the environment physically modifies part of the machine (its *receptor*).

2. The empirical information thus conveyed to the receptor is represented in machine language according to a program present in the receptor, producing an *initial classification* of this information in a set of *observation sentences* which form the linguistic input to the rest of the machine. We may assume at this stage that the initial classification represents a *loss of information* with respect to the empirical input, in two ways. First, however large the stock of predicate variables in the receptor's program, so long as it remains finite there will always be further observational respects in terms of which the objects *could* be described, but for which the receptor has no names. Second, with regard to the predicates it does have,

once the initial classification has been made, we have necessarily lost information about the detailed circumstances of recording that given properties and relations belong to given objects. We will assume, however, that recording is correct on at least some, and probably most occasions, although it may not be possible to judge at the time from the initial classification which occasions these are. This assumption will be called the *correspondence postulate*. To speak of 'correctness' and 'correspondence' here of course commits us to some form of ontology with regard to the empirical reference of the observation sentences, but it must be emphasized at this point that we are assuming that we have no access to information about the empirical world other than the observation sentences produced by the receptor.

3. The initial classification is likely to be highly complex and unmanageable. But it is also likely to suggest some immediate simplifications of itself. For example, we may be interested in finding universal lawlike generalizations within the observation sentences, or our desire for an economical and coherent system of laws and theories may involve more elaborate considerations, such as requirements of symmetry, simplicity, analogy, conformity with certain *a priori* conditions or metaphysical postulates. Without some such *coherence conditions* it is clear that a world described by even a small number of predicates in all possible combinations is likely to become quickly unmanageable, and not even the most rigorous inductivist is likely to suggest that no further processing of the observation sentences should take place.

The function of the coherence conditions is to produce from the initial classification a 'best theory', or range of best theories, conforming optimally both to the initial classification and to these conditions, where 'optimally' is itself defined by the conditions. Of course such a theory is likely to be in conflict with at least part of the initial classification. For example, although most P_1's are recorded as being P_2's, some may be recorded as $\sim P_2$'s, apparently refuting the potential generalization 'All P_1's are P_2's'. At the point where it is found that a certain theoretical system well satisfying the coherence conditions only approximately fits the initial classification, what can be done about the remaining mismatches? We now suppose that the structure of the machine permits internal feedback loops for the adjustment of theory to observation sentences, as well as the usual external loop which allows comparison of predictive output and empirical input. The internal feedback loops make a direct comparison of the current best theory with the observation sentences, and there are three kinds of internal modification of the machine which may be exploited to deal with mismatches here. I shall call machines equipped to deal with at most the first, second, or third kind *learning machines of the first, second, or third kind*.

First kind. Reinvestigate the empirical input that has produced the mismatches, and perhaps some nonanomalous input that seems to be related to it. It may be possible without much disturbance to the receptor program to reclassify certain of the objects to fit the best theory better. Recording an object as having a property *P* may for example have been physically a matter of its degree of resemblance to other objects recorded as *P*. Degrees of resemblance are not transitive, and thresholds of degrees may be modified in the receptor for peripheral objects without seriously changing the physical conditions under which something is recorded as *P*. In a complex classification such a change of threshold may also affect other, previously nonanomalous, observation sentences, and may result in quite radical changes of the whole classification and the best theory.

Second kind. If reinvestigation does not reveal that small changes of threshold are sufficient to produce a best theory with fewer anomalies, it may be convenient to 'change the meaning of' *P* in order to retain some form of generalization regarding this predicate. This might be done for example by modifying the receptor so that it distinguishes two predicates P_1 and P_1', discriminated by whether they are associated with P_2 or not, or by extending the application of P_2 to all P_1's. Put thus crudely, the proposal sounds like an abandonment of empirical constraints in favour of a kind of conventionalism of theory, but there are possibilities of comparison of different theoretical systems and different corrections of the first and second kinds into which empirical considerations will enter. In judging the empirical content of such proposals it must not be forgotten that the notion of coherence of the classification and of the related theory rests on the correspondence postulate that *most* of the initial classification is to be retained in the best theory, so that modifications cannot be made indiscriminately. The essential point, however, is that it cannot be known *a priori* and independently of the coherence conditions *which* part of the initial classification will be retained in the next successive best theory.

Third kind. We may also conceive that the coherence conditions themelves may, usually as a last resort, be modified in the light of success and failure of the sequence of best theories in accounting for the available observation sentences, and in making succesful predictions. There seem to be a number of examples of this kind of modification in the history of science: abandonment of the postulate of circular notions of the heavenly bodies; rejection of the notion that some theoretical postulates such as Euclidean geometry or universal determinism, can be known *a priori*; adoption and later rejection of the mechanical philosophy as a necessary condition of scientific explanation; the postulate of reducibility of organic processes to physicochemical theories. Such varieties of coherence conditions in fact constitute the main subject matter of philosophical dispute regarding science, and it will now be my contention that the more usual disputes between different accounts of the structure of science are almost vacuous except insofar as they reflect different views about the nature of the coherence conditions.

II

In order to justify this contention let us consider how different models of the structure of science now reveal themselves in various ways as simplifications of the structure of our learning machine. First, most classic models postulate a *stable observation language* (SOL),[2] that is to say, they make no allowance for correction of some subset of observation sentences, usually because of a tacit belief that unless some definite subset is incorrigible and known to be incorrigible in truth-value, or at least in meaning, no empirical check on the learning process is possible. We have seen that this belief is mistaken both with regard to truth-value and meaning. However, stability of truth-value and meaning of the observation language does not appear to be an essential postulate for the classic models of science, among which we may distinguish in particular *deductivism, inductivism*, and *conventionalism*. These models are characterized rather by their varying accounts of the nature of the coherence conditions, that is, of how the 'best theory' is arrived at.

Deductivism has not been very explicit about its coherence conditions. In its purest form it has held only that there is no inference or logic except deductive inference, and hence that the only definite conclusions that can be drawn in science are that some theories are falsified by empirical evidence by *modus tollens* inference. This view, it should be noted, is committed to SOL, since it has no means of choosing among as yet unfalsified theories, and hence no means of selecting a best theory in terms of which to distinguish acceptable and unacceptable observation sentences. However, some impure forms of deductivism have added to the falsifiability criterion some further criteria for grading theories not yet falsified. Among suggested criteria are, first, that theories should be *simple*. If this notion were spelled out in sufficient detail, it would provide for correction of observation sentences. Unfortunately, however, no fully satisfactory explication of simplicity is yet available, indeed it is very likely that this notion is ambiguous as between several different and partially conflicting explications, between which optimization criteria would have to be worked out. Second, theories are sometimes required to be *powerful*, that is to have maximal empirical content (entailments) beyond the available observation sentences, and hence to have maximal predictive power. This criterion too would provide the means of correcting some observation sentences, but as the sole criterion for theory-choice it is quite inadequate, since any theory may be made more powerful by conjoining to it some as yet unfalsified postulates unconnected with existing evidence. Hence the notion of 'most powerful theory' is empty.[3]

In the absence of clear notions of 'simplicity' and 'power', and of the optimal relation between them, deductivism does not have a definite account of theory-choice and theory-change. In deductivism questions of choice and change have in fact usually been relegated to a-logical discovery procedures, which are held to be historically, socially, and psychologically conditioned, but not to be the subject matter of philosophical analysis. *Inductivism*, on the other hand, has regarded questions of inference from evidence to theory, and hence of theory-choice and theory-change, as coming within the context of justification and as being susceptible to logical analysis. In the seventeenth-century version of inductivism, the postulate of the stable observation language was supplemented by the further belief that conclusions of inductive inference could be established as true rather than as merely probable, but present day inductivism has held that inductive conclusions can be graded only by probability values. It is my opinion, which I have no space to justify here, that in spite of the many objections which can be brought against such a probability model of induction, it is possible to develop a consistent theory of inductive confirmation with at least as much logical adequacy and rigour, as has been the case in deductivism. Among the coherence conditions in this view will be some constraints upon the prior probabilities of theories, and some rule such as Bayes theorem according to which probability distributions over theories are modified in the light of evidence. Probabilistic inductivism then provides an account of theory-choice and theory-change according to some such rule as: If a hitherto acceptable best theory is refuted or made increasingly improbable by accumulating evidence together with the coherence conditions, then change to a theory that is highly probable and that becomes increasingly probable by accumulating evidence and the coherence conditions. Unlike its seventeenth-century predecessor, this probabilistic version of inductivism can incorporate in its coherence conditions criteria for modification of some of the observation sentences for optimal fit with the best theory, hence this inductive model does not presuppose SOL.[4]

The third classically recognized model of science is *conventionalism*. This view may be held in a weak or a strong version. The strong version, when held in conjunction with SOL, would suppose that any systematic theory could be made to fit any empirical evidence, that is, no empirical input can result in the grading of theories. It would follow that no reliable predictions could be made either, since for every theory that made a specific prediction there would be another, not gradable as better or worse, which would make the contrary prediction. This strong version of conventionalism has, however, probably not been maintained by anybody, even Poincaré, for all so-called conventionalists who accept SOL have adopted some weaker version, according to which, while *some* theories are excluded by evidence, there remains a wide variety of acceptable theories which appear inconsistent with each other, but whose truth-values are not empirically decidable. Thus Poincaré claimed that any physical geometry can be fitted to any facts, so long as compensating modifications were made in mechanics and optics, and Quine asserts that 'Any statement can be held true come what may, *if* we make drastic adjustments elsewhere in the system' (my italics).[5] Such a weak version of conventionalism allows for some modification of the observation sentences as demanded by a given acceptable theory, but in the absence of some means of grading acceptable theories it is not clear how much modification can be allowed without cutting the empirical link altogether and reverting to strong conventionalism. Simplicity criteria are usually suggested as the source of such grading, in which case conventionalism without SOL may become indistinguishable from deductivism or inductivism without SOL. On the other hand, if SOL is maintained, weak conventionalism may be left with a subset of logically possible theories which are acceptable, but empirically ungradable with respect to that observation language, which seems to have been Poincaré's position.

By what criteria can we choose between these alternative models in their many varieties? Historically, there are many examples of attempts to determine *a priori* what general coherence conditions scientific theories must conform to. Such *a priori* approaches have been largely discredited, not only by philosophical argument, but also by counter-examples to many of these suggested conditions actually occurring in the historical development of scientific theories. At present there seem to be broadly three viable sources of coherence conditions, which may be designated *normative, innate,* and *value-determined.*

Normative studies are those most frequently found in analytical philosophy of science. They include most of the logic of deduction, confirmation and decision theory as applied to scientific theories, and also logical studies of simplicity, analogy, and classification. Their aim is to discover logical interrelations between various principles which seem intuitively to be requirements of good theories, so as to provide explications in logically consistent and economical models of science. In such studies failure of one of the models satisfactorily to explicate some intuitive requirement such as simplicity or predictivity usually counts against the adequacy of that model. The further question why a whole body of intuitive requirements are important and hence why a satisfactory model should incorporate them is held to be ultimately unanswerable, because it reduces to the inductive question: What are the most successful conditions for learning? And this question cannot be answered independently of the procedures of the learning machine itself. Indeed its unanswerability is the reason for allowing for third-kind modifications of the coherence conditions themselves as part of the learning process.

Recent experience with such normative studies discouragingly suggests, however, that they are not only incompetent to deal with the ultimate inductive problem, which is to be expected, but also that the kinds of constraints on coherence conditions suggested by purely logical or formal considerations are not nearly powerful enough to account for the comparatively economical and manageable character of most of our cognitive systems, or the comparatively limited mechanisms of data processing and storage that we have available, either in heads or hardware. It is likely that at least some of the more powerful coherence conditions that must be involved in scientific inference are *innate*, in the sense in which in transformational grammar deep structures of language are held to be innate. That is to say they are a genetic inheritance, perhaps selected during the course of evolution of learning organisms. To discover what such physically innate principles are, the philosopher of science must investigate such empirical disciplines as taxonomy, cognitive psychology, artificial intelligence, empirical decision-making, and structural linguistics. Little has yet been done to bring together relevant findings in these fields to illuminate conditions of scientific learning.

What I have designated *value-determined* coherence conditions may be exemplified in the past history of the physical sciences, and in the present state of softer sciences. Such conditions would include theological objections to the heliocentric system, and moral objections to the theory of natural selection, to mechanical theories of the mind, and to theories of social determinism. It is commonly supposed that physics has outgrown such conditions, and that the soft sciences, in becoming hard, will outgrow them in their turn. It is also possible, however, and perhaps more likely, that the way we now delimit the physical sciences is just constituted by those areas of experience which can be relatively freed from value-determined conditions, and that the sciences of organisms and particularly of man cannot be so freed without disturbing effects on the values man places on himself and his society. In the latter case, the decision to pursue value-free science would itself be a value-decision in the social domain. If this argument is found unacceptably ideological, however, another consideration is relevant. The coherence conditions of even the physical sciences have been found to be severely underdetermined by normative considerations and the pragmatic learning process. It is very likely that the conditions of sciences of more complex systems will be found so severely underdetermined as to demand a more stringent framework, and in the absence in our culture of theological or metaphysical constraints, adoption of social and ethical constraints cannot be ruled out.

There looms here, of course, the vexing question of the nature of scientific *truth*. It arises most pressingly in light of the value considerations just advanced, but it should be noticed that it is also present, though less disturbingly, in the whole analysis of scientific learning that has been implicit here. In the rest of this paper I shall take up the question of what are the consequences for scientific *realism* of the classic models of science I have described in terms of the learning machine, or, what amounts to the same thing, the question whether truth-value can be assigned in principle to sentences of the theoretical system produced by the machine.

III

First, if we are given an observation language stable in meaning, then the fact that the observation sentences have empirical reference and hence truth-value has been held to be unproblematic. But the question has been asked, 'Do the theoretical sentences have truth-value?' From the deductivist point of view they need not. This has been argued by Nagel[6] in his demonstration that within the limits of deductivism no distinction can be made between the realist and instrumentalist interpretations of theories, which is the same as to say that an instrumentalist interpretation is sufficient. However, theories may without inconsistency be given realistic interpretation in deductivism, but to do so is to add an extra postulate not required by the logic of deductivism itself. Probabilistic inductivism, on the other hand, would seem to be committed to realism, since on that view theoretical sentences are in some sense the consequences of observation sentences, which do have truth- or probability-values, together with rules of probabilistic inference. Unless the interpretation of probability here is wholly subjective, to assert that a sentence has probability-value is to assert that it has a probability of being *true*, and with respect to the probability of theories, even the most convinced subjectivist can hardly fail to hold that his subjective degree of belief in a theory, if rational, is a degree of belief in its truth, on which, in some circumstances, he would be prepared to act.

If the stable observation language is rejected, however, this rejection will have a profound effect on the conception of truth appropriate to scientific theories. Even in the learning machine of the first kind, the truth of individual sentences becomes a function of coherence conditions as well as of empirical input in such a way that, although all observation sentences are assumed to have truth-value, we cannot assign such a value independently of the coherence conditions. In the learning machine of the second kind meaning also is relative to both coherence conditions and empirical input, and hence the reference or correspondence element of meaning of even the observation sentences becomes problematic. However, if we remember the conditions under which the learning machine is said to be self-correcting, the realistic reference of observation sentences is guaranteed in the minimal sense that some proportion of them is asserted to be true, that is to say, true as representations of the empirical input in that language which is for the time being programmed into the receptor. With this understanding of the realistic reference of at least a subset of sentences of the system, the implications of deductivism and inductivism for the realism of theories would seem to be the same as before. Deductivist accounts which depend only on a simplicity grading of theories require no more than an instrumentalist interpretation, but are consistent with realism as an extra postulate; inductivist accounts presuppose realism of all parts of the theoretical system, since all sentences are assigned probability values.

It should be noticed, however, that a realism which is consistent with either of these models of science is very different from the sort of naïve cumulative realism that might have been presupposed by a seventeenth-century inductivist. This is the view that scientific theory is a continuously progressive, cumulative and convergent approach to truth, where truth is understood as correspondence between systems of theoretical sentences and the world. The history of science has already sufficiently demonstrated that successive acceptable theories are often in radical conceptual contradiction with each other. The succession of theories of the atom, for example, exhibits no 'convergence' in descriptions of the nature of fundamental particles, but oscillates between continuity and discontinuity, field conceptions and particle conceptions, and even speculatively among different topologies of space. But even apart from the historical evidence, neither deductivism nor probabilistic inductivism offers any grounds for such a cumulative realism. Deductivism is quite consistent with a historical sequence of radically different theories which correspond only in entailing certain observation sentences in common, and if SOL is abandoned, these observation sentences need only be in common between pairs of historically successive theories. In the case of probabilistic inductivism each successive theory has some probability of being true given the current evidence and current empirical reference of terms, but there is no guarantee that these theories will exhibit any conceptual convergence.

Rejection of the stable observation language has, however, often been associated with a more radical departure from the classic models of science. This is a view which may somewhat misleadingly be called *idealism*, but more accurately *pluralism*. It amounts to a strong conventionalism without SOL, since it asserts that any theoretical system can be accepted as scientific, unconstrained by evidence. It is, however, more plausible than strong conventionalism with SOL, because it exploits the absence of stability to show how any given theoretical system so permeates the putative observation sentences that it transforms them into true sentences of the system. In terms of the learning machine this view would represent the observation sentences as deriving the whole of their meaning from the theoretical system as defined by the coherence conditions, and not at all from the empirical input. In this case no sentence can even be said to be *false*, for, since every sentence of the system is true in virtue of the meaning it derives from the system, a sentence which contradicts the system is necessarily outside the system and hence is not false but meaningless. In this respect the view is reminiscent of classic idealism; however, it differs in that classic idealism is *monistic*, whereas this view is *pluralistic*. Classic idealism would be represented in the learning machine by the claim that the coherence conditions uniquely define a single logically consistent theoretical system, which determines the meaning and truth of each sentence of the system independently of test by empirical input, but which must, because it is unique, nevertheless be true also of the totality of the world insofar as that world is susceptible of description at all.[7] Pluralism, on the other hand, is the claim that though coherence conditions wholly determine meaning and truth, neither the set of conditions nor the theoretical system is unique. Pluralism in this sense clearly describes no learning mechanism with respect to the empirical world, and is from the present point of view quite uninteresting, since it is the assertion that there are no constraints on the learning machine, either on empirical input or on the coherence conditions. Pluralism also clearly excludes a realistic interpretation of theories, since no grounds for realism are given either in the truth-relation of some observation sentences with the empirical, or in a claimed uniqueness of theoretical system which guarantees its truth of the totality of the world.

Reverting to those models which do permit realism of a kind, it may seem strange to speak of a realistic interpretation of theories unless there is some sense in which the succession of theories can be said to add to an accumulating deposit of fact.[8] Unfortunately it does not appear to be the case that such an accumulating deposit can be guaranteed merely in virtue of the nature of science as a learning process, or made part of the conditions for an enterprise to be scientific. If science is essentially a learning process, its aim is efficiency of successful learning, and the learning machine that has been described is one which, as far as we can tell in our kind of environment, will continue to learn. But we have no idea what is the most efficient method of learning. It may be, as a mild form of pluralism might suggest, that the quickest learning will take place by frequent and radical changes both of theories and coherence conditions. Or it may be that insertion of some randomness into the machine's selection of best theories and predictions may be advantageous.

In fact, however, scrutiny of the history of science reveals somewhat more continuity than is logically guaranteed. The form of laws, or approximations to laws, have generally been maintained throughout successive theories, and so have the classifications of what count as similar systems implied by these laws. For example, that the earth, and stones falling on the earth, satisfy the same laws as the other planets, has been maintained through the relativistic revolution, and so have the approximate forms of these laws within certain empirical limits. These facts are not affected by conceptual changes in the theory of space and time, or in the understanding of mass and its natural motions, which affect all these bodies alike. Similarly, that unobservable gas molecules satisfy approximately the same laws as macroscopic particles is unaffected by changes in the theoretical conception of 'particle', which affects both microscopic and macroscopic particles alike. Facts of lawlike structure and similarities of nature between physical systems seem to be maintained and to be cumulative. Theoretical interpretations of what the natures of these systems absolutely are, are not. But in view of the relativity of language to the receptor's program, 'what anything absolutely is' is a chimera, born of illicit reification of language. The persistence of structures of laws and similarities of systems is not inconsistent with realism however, for they are real structures and real similarities. They are, in fact, what replace the postulate of the stable observation language, although as I have argued, even their persistence is not strictly necessary to the learning process which constitutes science any more than the stable observation language has been found to be.

I have presupposed throughout this paper that the purpose of science is to learn truths about the empirical world, and that such learning is subject to pragmatic tests. If there is to be any demarcation of science as publicly, and generally speaking historically, understood, then it must be in terms of learning and empiricism, or as it used to be put, in terms of an empirical epistemology. This epistemology is, I claim, exhaustively represented by the learning machine. Whether the machine also exhaustively represents the *ontology* of science is the question of whether there are sources of truth for the coherence conditions other than the logical and the pragmatic, which is the still unanswered question bequeathed to us by positivism, and it is unlikely to be answerable in the context of empirical science alone.

Notes

1 For illuminating discussions of hierarchically organized learning machines I am indebted to James S. Williamsen, cf. his 'Induction and artificial intelligence', unpublished Ph.D. thesis, Cambridge, 1970.

2 This postulate was first drawn to attention in a way significant for current philosophy of science in P. K. Feyerabend, 'An attempt at a realistic interpretation of experience', *Proc. Arist. Soc.* vol. lviii, 1958, 143–70.

3 An influential attempt to relate 'simplicity' and 'power' is to be found in K. R. Popper, *The Logic of Scientific Discovery*, London, 1959, ch. 7.
4 I have discussed the possibility of correcting the observation sentences in a probabalistic theory in M. Hesse, 'A self-correcting observation language' in *Logic Methodology and Philosphy of Science*, vol. iii, ed. B. van Rootselaar and J. F. Staal, Amsterdam, 1968, 297–309.
5 H. Poincaré, *Science and Hypothesis*, English trans., New York, 1905, Part II; W. v. O. Quine, *Two Dogmas of Empiricism: from a logical point of view*, Cambridge, Mass., 1953, p. 43.
6 E. Nagel, *The Structure of Science*, London, 1961, ch. 6.
7 It may seem paradoxical to imply here that classic idealism is realistic. I do so only in the sense that in that view coherence conditions of truth must *ipso facto* be true of the world if anything is, not in the sense that truth is a relation of correspondence with the world.
8 Cf. the discussion in G. Buchdahl, 'Is science cumulative? Some ontological explorations, *New Edinburgh Review*, 1971, p. 4.

6 *Truth and the Growth of Scientific Knowledge*

This chapter is a reaction to two themes in recent history and philosophy of science. The first is the revival in philosophy of language of interest in truth and meaning, which relates to problems long discussed in philosophy of science. My tentative conclusion here is that tools are now available for the solution of these traditional problems, but that the detailed discussion by philosophers of language tends to bypass philosophy of science, chiefly because it does not take sufficiently seriously questions about truth and meaning in theories and above all the question of radical theory change. A second theme, which is more implicit than explicit here, concerns the fashionable tendency in post-Kuhnian philosophy of science for discussing philosophical problems in terms of case histories. These discussions have been motivated by the desire to avoid the threat of relativism and to restore rationality to theoretical science. Part of my argument here is intended to show that case histories alone are inadequate for this task, and that a certain division of labour between historians and philosophers of science needs to be restored.

I begin by accepting two principles drawn from recent discussions, namely a certain interpretation of theory–observation continuity, and a principle of no privilege for the theoretical ontologies of our science relative to other scientific ontologies. Then I develop criteria for an *epistemological* theory of truth for theoretical sentences in terms of (1.) a consensus theory of truth for some observation sentences, (2.) a probability theory of degrees of belief for theoretical sentences, (3.) a principle of charitable translation for alien scientific systems, and (4.) a principle of scientific growth. These criteria are, I argue, sufficient to define a univocal concept of truth for science, and are sufficient for scientific realism. Finally, I make some comments on the tasks of the historian and the philosopher of science in relation to these problems.

1. Theories of Truth

There are two familiar propositions to be considered about the truth of scientific theories:

A. *Instrumentalism:* As old-fashioned instrumentalists used to hold, no theoretical sentences have reference or truth value.

B. *Theory–observation continuity:* As is cogently argued in more recent literature, there is no sharp logical distinction between theoretical and observation sentences, implying at least that either all have truth value or that none do.

If we accept both *A* and *B*, then there are no descriptive sentences in science, including 'observation sentences', that have reference or truth value. This is a position implied by extreme 'meaning variance' theorists (for example Feyerabend sometimes), according to whom 'truth', if it is a useful concept at all, has to be 'truth within a system', that is, sentences are 'true' in virtue of inference from the premisses of the system (and even this version of the theory is weaker than some in presupposing no serious variance in the nature of the inference rules from one theoretical sentence to another). Since it follows in this view that even so-called singular observation sentences like 'this leaf is green' can be given truth value only in this non-referential sense, I shall assume it is too counter-intuitive to be acceptable. Indeed the relevance of theories of truth and reference to theoretical science arises just because rejection of this view leaves the status of scientific truth unclear.

With regard to *B*, it will turn out later in this paper that it is necessary to reinstate some forms of distinction between theoretical and observation sentences, if only distinctions of degree. But let us for purposes of investigation of the truth claims of theories accept that there is no *sharp* distinction such as would be implied by ascribing truth value to one set of sentences and not the other. This, together with the need to ascribe truth value to observation sentences, implies rejection of *A*. Both observation and theoretical sentences have truth value, the question is, how do they have it?

Recent discussions of truth have taken Tarski's theory as their starting point. However, I am going to suggest that with regard to the problem of truth for theoretical sentences, and even for observation sentences, Tarski's theory is at best incomplete. First I shall accept a point most clearly made recently by Putnam, that though the *form* of Tarski's theory may be a necessary framework for talking about 'truth' or some necessary surrogate of 'truth', the interpretation of the theory is not given uniquely by its form. In particular as Putnam shows,[1] we can replace the classic T-sentence T1: '"Snow is white" is *true* if and only if snow is white' by using a set of intuitionistically defined connectives such that '"Snow is white" is *warrantably asserted* if and only if "Snow is white" is *provable* in the relevant theory'—a reformulation which would give a non-standard sense of 'truth' suitable for meaning variance theorists who understand 'truth' only as coherence within a system. Thus it is only in a 'standard' interpretation using classical propositional logic that Tarski's theory provides a theory of truth *and reference*. As a formal metalinguistic theory alone Tarski's theory is not sufficient to explicate truth in a correspondence sense.

Secondly, I shall accept a point made by Hartry Field[2] that Tarski's theory alone fails to effect a reduction of the semantic notion of 'truth' to purely syntactic and thence to *physicalist* notions (which Field claims plausibly was Tarski's programme, consistently with the surrounding physicalism of 1930s philosophy of science) The theory in fact leaves further implicit semantic notions to be elucidated. This is most evident in the case of the theoretical sentences we are considering here, for example suppose there is a Tarski T-sentence T2: '"Electrons are round" is true if and only if S', where S is a sentence in the metalanguage. What sentence? Well, if 'S' is 'electrons are round', by analogy with the paradigm case T1, T2 becomes entirely inapplicable as an explication of truth. If we are worried about the truth and reference of 'electrons are round' in the object language, we are equally worried about S in the metalanguage. At least in T1 we may perhaps regard the occurrence of 'snow is white' in the metalanguage as serving to point us to a 'physicalist' situation of the sort to which, as Field argues, Tarski was trying to reduce the notion of truth. But in the analogous substitution of S in T2 this is not so—our problem has arisen precisely because we do *not* know how to construe the physicalist concomitants of 'electrons are round'. The attempts of Carnap and others to reduce 'electrons are round' to an equivalent observation sentence correspond to a second suggestion for the replacement of S, namely that it be such an equivalent observation sentence. If these attempts had been successful as explications of what 'electrons are round' means in physics, they would have shown how to extend Tarski's theory to theoretical sentences. But these attempts conclusively failed. It follows that without supplementation Tarski's theory of truth is entirely uninformative about the specific problem of the truth and reference of theoretical sentences.

If Field's argument is correct, however, it is not just a question of finding supplementary explications for *theoretical* sentences, because the explication it provides for *all* sentences is insufficient. What is intended to be the 'pointing to' a physical situation performed by the occurrence of an observation sentence like 'snow is white' (or 'this snow is white') in the metalanguage, is itself in need of semantic explication, for at best it only dispenses with the notion of 'truth' at the cost of presupposing that we understand other semantic notions such as 'satisfies', 'denotes', 'applies to'. In other words, in Tarski's theory it is assumed that the understanding of truth and reference is given in the metalanguage for all sentences alike, whereas debates about the truth and reference of theoretical sentences in science have shown that we have no such metalinguistic account. Moreover, attempts to find such an account have emphasized that no distinction can be drawn between theoretical and observation sentences in these respects—both are problematic, and in both cases we need something more like a causal story to indicate how a language community comes to utter true and meaningful descriptive sentences.[3] These discussions have also emphasized that theory change has to be taken as normative for the understanding of theoretical meaning, and have led to the much debated problems of meaning variance and incommensurability.

I shall adopt from these debates what I call a *principle*

of no privilege, according to which our own scientific theories are held to be as much subject to radical conceptual change as past theories are seen to be. Insofar as Tarski's theory and those based on it are taken to presuppose the unproblematic givenness of something called 'our science', they clearly violate this principle. Moreover the principle entails that any supplementation of a theory of truth to explicate the relation between world and language that is glossed by the concept of 'satisfaction', must also take account of the dynamic character of theories and theoretical meanings. Kripke's influential theory of reference seems to violate the principle of no privilege in both these respects, insofar as it assumes our science is privileged, and holds that designation is rigid. Consequently it is in this respect an inadequate supplement to Tarski's theory. Quine, on the other hand, has developed a theory which both takes account of the essentially causal character of observational meaning, and also allows flexibility in the relation between theoretical truth and theoretical meaning. It is the suggestions implicit in 'Two Dogmas' and part of *Word and Object*[4] that I shall pursue, rather than those of Quine's later work, because it is the doctrine of underdetermination of theories rather than that of indeterminacy of translation that appears more relevant to theoretical science.

2. Observation sentences

Let us begin again with the problem of the observation language. The dilemma is this. First we do not want to assume with the old positivism that there is a deposit of observation sentences that retain constant and unproblematic meaning and truth value throughout theory change. On the other hand we wish to retain the possibility that at least some observation terms associated with each theory have extensional references and that some observation sentences are true. I shall later introduce the notion of the *probability* of observation and theoretical sentences, but it should be noted here that this will be understood primarily in an *epistemological* sense, and will not do as a replacement of the objective concept of truth. For if probabilities were to be used (objectively and ontologically) as a replacement for truth, the only objective concept of probability that seems appropriate would be that of a frequency of truth among all observation sentences ('probably true' cannot here mean 'somewhere between true and false' for each separate observation sentence). In this case we still have to analyze the notion of some observation sentences being true, though it does not require that we be able to identify exactly which these are. A second difficulty is that if we measure the probability of any given sentence conditionally upon some evidence, then the probability of the evidence sentence conditionally upon itself in the new posterior distribution is identically 1, that is, the evidence sentence itself must be taken as true in the conditional distribution. We cannot therefore bypass the problem of truth, at least of some observation sentences, by speaking of their probability.

Since we are adopting a principle of no privilege, we can consider any single theory, or group of theories formulated in a single natural and theoretical language, and consider how a subset of the sentences (the 'observation sentences') of that language can be said to have reference and truth in relation to the world outside the theory. There is a minimal answer agreed by almost everyone, namely that for a given language community, 'true' observation sentences and 'correct' application of general observation terms are at least those that are reinforced as such by the consensus of the community. A theory of reference that starts there might be called a 'consensus' theory, as distinct from the traditional 'coherence' and 'correspondence' theories, but of course more than mere consensus is implied by inclusion of the word 'observation' in its specification. It does not follow in such a theory that everything that is agreed is true, nor that truth is wholly dependent on the language community. Such objections rest on a misunderstanding, for the mechanics of language learning and reinforcement of 'correctness' *themselves* depend on external reference of language. It is not that anything goes if agreed by the language community, but that the language community does or does not agree according to external constraints. 'Grass is green', not because there is some extra-linguistic universal 'greenness' that is intuitively and universally recognized, but also not because there is a capricious and arbitrary agreement that the sentence should be true. 'Grass is green' because that is the way all recognizably similar objects and colours have been acceptably described in English. If grass turns brown the consensus will cease to describe it as green, because that is the way we have learned, or have evolved, to use our observational vocabulary.

How, then, does the consensual truth of observation sentences in one language community relate to those of another language community? Quine's answer in *Word and Object* was that the set of singular observational 'occasion sentences' goes over with unproblematic truth value into our occasion sentences, thus providing some of the fixed points of the translation manual. His subsequent answer implied in 'Natural Kinds'[5] and other papers seems to be that, since we now have a good science of real kinds, we can identify for certain what others were really talking about and translate accordingly. This second assumption is certainly stronger than is required by the problem, and even the first assumption of unproblematic translation into our occasion sentences is too strong also. There is no need to assume any 'fixed points' of translation, but only that, according to the *principle of charity* discussed by Quine, Davidson and others[6] we try to translate the sentences of any alien language, particularly the observational ones, in such a way as to make as many as possible come out true in our language. This process need not be assumed to have unique solution, nor need it entail any sharp distinction between theoretical and observation sentences.

We assume, then, that we have sets of singular observation sentences whose reference and truth value are given in the consensus of their respective language communities. The set S1 in language L1 goes over into a set S2 in L2 under some set of translation rules which as far as possible preserve truth value. I shall not say more about the nature of these translation rules at present, partly because there are detachable problems here for the philosophers of language to solve, and partly also because I do not regard the translation process as entirely independent of the problem of theoretical meaning, and we have yet to consider how to extend the analysis of truth and meaning to theories.

3. Theoretical sentences

The problem about theories is this. Every scientific system implies a conceptual classification of the world into an ontology of fundamental entities and properties—it is an attempt to answer the question 'What is the world really made of?' But it is exactly these ontologies that are most subject to radical change throughout the history of science. Therefore in the spirit of the principle of no privilege, it seems that we must say either that all these ontologies are true, i.e.: we must give a realistic interpretation of all of them, or we must say they are all false. But they cannot all be true in the same world, because they contain conflicting answers to the question 'What is the world made of?' Therefore they must all be false. But if they are all false, the notion of a realistic interpretation of theories seems redundant—realism surely implies that *sometimes* theories are true, and if they are never true, instrumentalism seems a far more natural interpretation. In what sense can we retain some theoretical truth while accepting no privilege?

One way of explicating this requirement is to retain the conclusion that some theory is true, but weaken one of the common implications of realism, namely that we can sometimes *know* that we have true theories. We can than weaken the demand that we know that some specified theoretical sentences are true, to a request that we know some probabilities of truth short of certainty. Probability will be interpreted here in two ways, one ontological and one epistemological. In the ontological sense we follow the logic of the above argument and accept that all universally quantified sentences ascribing basic ontology to the world have probability zero, that is, are almost certainly false. At best some subset of these may actually be true, though we can never know that they are; at worst all that is true is the negation of some disjunction of universal sentences, which is fairly uninformative. However, it does not follow that all theoretical sentences in *finite* domains are false; some may be false and some true. The probability of a theoretical sentence in a finite domain may therefore be interpreted as a frequency of true sentences describing individuals in the finite domain of such individuals. More usefully, however, probability has an epistemological aspect. In this sense the probability of a sentence is to be interpreted as a rational or warranted degree of belief of the relevant scientific community that, conditionally on the total known evidence, the sentence in question is true. This epistemological interpretation is quite consistent with the view that all sentences actually have definite truth value (except perhaps for some sentences in quantum theory, but that is a different question), but it allows for the fact that definite truth values cannot generally be asserted. Rather, sentences are proposed with comparatively lower or higher degrees of belief in their truth. This is intended to explicate the intersubjective judgments of scientists that one theory is better confirmed by evidence than another, or that it is almost certainly refuted, or that it is *a priori* plausible in the light of background knowledge, and so on. As has already been argued, it is necessary to identify some sentences of the system (observation sentences 'at the periphery' as Quine would say), as those which are accepted as true by a given language community.

The picture of the probability distribution over a Quinean theoretical network will then go something like this. At the periphery will be singular observation sentences of some kind, which are given probability 1. Lower level generalizations of which these observation sentences are substitution instances have probability, conditionally upon these sentences, less than 1. Consistently with the probability axioms, in general the broader the generalization over a given domain the lower the probability. Inference relations, which take us from the periphery to the highest level theoretical generalizations at the core of the network, are not only inverses of deductive inferences, as in the DN model, but also include whatever direct inductive links may be acceptable, including statistical arguments, and arguments from *a priori* preferences between hypotheses, for example in terms of their simplicity or analogy with other highly probable hypotheses. At any given stage of development of the theoretical system we assume there is some probability distribution over its sentences, specified at least by comparative probability values, not necessarily by exact numerical values, since these could not be other than artificial as measures of degrees of belief. This distribution is the outcome of past experimental evidence and past judgments of the *a priori* plausibility of various kinds of hypotheses. It is modified both by more evidence at the periphery, and possibly also by 'shifts of opinion' about hypotheses, represented by changes in their prior probability. The most familiar method of modifying such conditional probability distributions is by the operation of Bayes theorem, but other methods have also been suggested as better explications of particular kinds of scientific inference.

The detailed working out of such a picture of theoretical inference is not to the point here.[7] However, one feature in which it appears to differ from Quine's model should be mentioned; that is the question of *entrenchment*. Quine supposed that theoretical sentences 'near the core' would be the last to be given up in any modification of theory originating from the empirical periphery: for example conservation of energy, or the principle of local causality. It looks as though such sentences ought to be given *high* probability in my model, whereas I have said that in general probabilities decline from periphery to core. The difficulty is only apparent, however. Quine's metaphor of the net can be taken most naturally to refer to a *single* theoretical system which is for the time being accepted, subject to constraints and modifications originating at the periphery. The core sentences are those very general theoretical sentences which determine the conceptual framework of a theory, and which are tested at the periphery only via a complex of inferences involving other less general sentences, boundary conditions, etc. They can therefore be regarded as entrenched relative to these less general associated sentences. In my model, on the other hand, the network consists of *all* well-formed sentences of the language, over which there is a probability distribution whose values are related by deductive and inductive rules consistently with the probability axioms. Thus Quine's entrenched sentences will be given high probability (will be highly confirmed) relative to their own prior probabilities and conditionally upon all the evidence, and in particular higher (perhaps much higher) probability than their theoretical rivals—those sentences inconsistent with them which appear in my network but not in Quine's. In the light of the evident occurrence of radical theory change, however, it is reasonable to suppose that no non-analytic theoretical sentence however entrenched is ever assigned probability 1.

We can now try to adapt the probability model to the translation problem for other people's theoretical systems. It may be suggested that this is just a matter of straightforward application of the principle of charity; that is, adopt that translation manual that makes as many as possible of their sentences come out true in our theory. This, however, is clearly *not* just what we want in the case of alien scientific systems, because we may want to say that although most of their observation sentences were true, a great deal of the theory of, say Aristotle or Descartes, was false. There are two ways in which we might understand this. First we may say it is false because it contradicts our theory. But this is unsatisfactory, both because it violates the principle of no privilege, and also because it begs the question of good translation while trying to specify what the criteria of good translation are: before we have applied the criteria, how do we know their theories come out false in ours? But we may understand our intuition that their theories are false in another way, namely that they are not given high probability by good scientific inferences from *their* observation sentences. In practice this will come out looking like a judgment that their theories are false by comparison with ours assumed true, but this is not because of direct comparison but rather because their inference patterns do not have the same structure as ours. It may be objected that this is to give privilege to our inferences if not to our theories. Indeed it does, but then we are talking about alien *scientific* systems, not just any alien belief systems, and we are surely entitled to recognize as scientific only those systems that can be seen to be subject to our criteria of observational test and the inference patterns that lead to predictive success. This is a prescription, not a description, and therefore it is not upset by the revolutionary induction from the history of science about theory change. (I shall return to this point in the last section of this chapter.)

The procedure for translating alien scientific systems will then go something like this. Select some subset of the sentences that appear in our interaction with the alien system to correspond to our true observation sentences, and give these high initial probability. In the case of history of science not too far removed from our own (for example, Joseph Priestley's investigations which are used as illustrations below) relevant interaction with them may include not only manuscript evidence, but also discovery or reconstruction of the actual objects and apparatus constituting their empirical evidence. Then the probability of one of their theoretical sentences (say T_1) being true, given their evidence (say E_1), can be understood in the same way as the probability of our theoretical sentences, that is in terms of the rules of inductive inference expressed in a conditional probability theory. The probability of T_1 given E_1 as expressed in our observational translation E_2 is then the product of the probability of T_1 given E_1, and the probability of E_1 given E_2. The principle of charity can then be modified to read: adopt that translation manual that gives their theories (both theoretical and observation sentences) as high a joint probability as possible.

Naturally, the solution cannot be expected to be unique, if only because the judgments of probability values are necessarily imprecise. But this situation corresponds to the ever-present possibility of alternative historical interpretations of alien scientific systems. Also, of course, it must be emphasized that it would inappropriate to suggest that the *historian* engages in the explicit probability guesswork I have just described, or that he should suppose that past scientists actually judged their theories in this way. In the first place, the historian may rightly be more interested in the coherence of the alien system in its own terms than in interpreting its truth in our terms, and in the second place, as all philosophers of science know only too well, even contemporary scientists are reluctant to recognize explicitly their own implicit rules of inductive inference. The interest of the procedure is a *philosophical* one: it is to give a sense of the truth values of alien belief systems where these are understood to be scientific in the same sense that ours are scientific.

The modified principle of charity permits certain further refinements in the interpretation of truth and meaning.[8] The procedure I have just described does not give the absolute probability of the 'best translation' of T_1 (say T_2) in our system, because we may be able to judge T_2 on evidence not available in the alien system. But the probability of T_1 in their system gives one of the criteria for best translation, on the charitable assumption that the alien system intended truth on its best evidence. In these terms, T_1 may be well inferred from E_1, but have low probability on our more comprehensive evidence. Or T_1 may be badly inferred from E_1, and therefore of low probability in their system, and may either have low probability also in ours, or it may be a 'lucky guess', unjustified by E_1, but turning out highly probable on our extra evidence. And in general, the fact that all the judgments are probabilistic except the judgments of some of our observation sentences based on consensus, means that we leave it open to future systems, possibly with more evidence, to modify both their and our scientific beliefs, that is to modify all the probabilities. The whole account depends crucially on recognition of some rules of inductive inference (an unpopular move with most philosophers of science at present). But without such rules it does not seem possible to give an adequate theory of translation for theoretical sentences.

Quine is one of the few philosophers who has recognized that induction has a place in this discussion. But he has argued that there is an important distinction to be made between the underdetermination of scientific theories, which he seems to think is in principle decidable by what he calls 'normal scientific induction'[9] and the indeterminancy of translation between alien belief-systems. Consequently, while he would agree that our scientific theories have truth values decided by inductive methods, he holds that there is no question of truth value between systems under radical translation. He has not spelled out in detail what he takes 'normal scientific induction' to be, but in view of his assertions to the effect that our science is the best account of the world we have it is almost certain that he underestimates the radical character of changes in scientific ontology. These seem to bring the present account of science nearer to what he calls 'ontological relativity', and hence to the domain of indeterminacy. But whatever may be the case for translations between non-scientific belief systems, there seems to be no case for holding that 'normal scientific induction', properly formulated, cannot cope with the problem of translation between different scientific ontologies along the lines described above, as well as the problem of comparing different theories within one scientific language.

The consensus theory of truth and the probability model give a univocal sense to the 'truth' of theoretical sentences that are not directly constrained by the external world, through the conditional probability of these sentences in the theoretical network. And since theoretical sentences can be said to have truth value, they can also be said to have reference in the same way that observation sentences have reference, that is, if they are true, there are entities and properties in the world as they describe them. Theoretical meaning is referential—it is given by the meaning and truth value of observation sentences together with the inferential rules that relate the truth value of theories to the truth values of observation sentences. Meaning is not, as in 'meaning variance' theories, given independently of observation constraint and purely by theoretical context. Thus within the observational vocabulary of a given language community no problems of meaning variance arise. However, we still have to consider in more detail how translation of theoretical sentences between language communities works, and to ask whether the number of true scientific sentences increases with time, that is, is there growth of scientific knowledge?

4. Growth of scientific knowledge

In his paper, 'What is "Realism"?'[10] Putnam has argued that without the assumption that there is some accumulation, progress, or convergence of scientific theories towards the truth, the *success* of science in correctly predicting observable phenomena would be a miracle. He therefore regards a realism that safeguards such convergence as an *explanation* of the success of science. 'Explanation' seems to be used here in a 'metaphysical' rather than a scientific sense, for Putnam does not claim that it is falsifiable by any empirical facts (for example, it would not be falsified if science ceased to give successful predictions, for that might mean only that our scientific theories were not good enough). Neither is realism *entailed* by scientific success, because science might after all be a miracle. But realism is part of a set of sufficient conditions for success, without which predictions and scientists' belief in science seems to be irrational.

Putnam goes on to argue that this kind of realism requires that we block the induction drawn from the history of science by the revolutionaries, that is the conclusion that there is no continuity or convergence of scientific theories, and that therefore they have no real reference. I cannot find any direct *argument* in Putnam's paper that this induction must be blocked; he seems rather to treat this as one of the conditions of realism, which is in turn one of the conditions for explaining scientific success. It is not clear, however, that we need such a strong version of realism in order to provide this explanation. There are two opposing constraints to be kept in mind here. The first is the principle of no privilege for our theory which arises from accepting the induction from the history of science. Putnam casually violated the principle by remarking in relation to what he calls the Quinean predicament: 'What other theory can *we* use but our own present theory? . . . Well, we should use *someone else's* conceptual system?' (pp. 182, 192). Secondly, there is the constraint of the *principle of growth*, according to which science does exhibit apparent cumulative predictive success. Can both of these principles be satisfied together?

According to the first principle it follows that if sentences in one scientific system have truth value, then sentences in all scientific systems have truth value in the same sense. We have seen in terms of the probability model that some 'peripheral' observation sentences and their translations can be held to be mostly true *independently* of the system. What, however, about the theoretical sentences? Putnam suggests two conditions for theory to converge towards true descriptions of the real:

1. Terms in mature science typically *refer*—that is,

although some turn out not to (for example, phlogiston), most do, (for example, electrons).

2. Laws in mature science are typically approximately true.

There are two problems about these formulations. First there is the possibility, emphasized by revolutionaries, that *all* our theoretical terms will, in the natural course of scientific development, share the demise of phlogiston. Secondly, if that is the case, a question arises about what can be meant by 'approximate' truth when there is discontinuity of reference.

Let us look more closely at the way in which 'phlogiston' and 'electron' may be said to refer. To say simply, as in Putnam's (1), that terms *typically* refer, with exceptions that turn out to be terms of superseded theories, is to give *prima facie* privilege to our theories. For what terms 'typical' of, say Priestley's phlogiston theory, *do* refer in the sense in which it is claimed that electron does? Do 'air', 'caloric', 'dephlogisticated air', 'fixed air' . . .? In all these cases, although there may be referring terms in our theory that bear some resemblance to Priestley's terms, it is not a close enough resemblance to make us want to say, Yes, clearly Priestley *meant* to refer to our oxygen, carbon dioxide, or whatever. To say this would be to beg all the questions illuminatingly raised about theoretical meaning in historical studies. To see what is going on without initial presuppositions of privilege it is better not to take an example from our own theories, so let us take the case of phlogiston.[11] Here are some typical assertions concerning phlogiston in eighteenth-century theory:

(*a*) When metal calcinates, phlogiston is given off.
(*b*) Particles of phlogiston repel matter particles.
(*c*) When oil of vitriol is poured over zinc, phlogiston is given off.
(*d*) Phlogiston is inflammable.

The first two, (*a*) and (*b*) are sentences used in the context of an experiment we should describe in terms of the oxydation of metal, with accompanying increase in weight. From our point of view, (*a*) is false, not only because we do not believe phlogiston exists, but because nothing is *given off*; (*b*) is false because there is no such thing as phlogiston. (I assume here that when a law is seriously proposed in a scientific theory, it must be taken to include the assertion either of the *existence* of the antecedent and consequent, or that the antecedent and consequent are *idealizations* of existents which are usefully defined within that theory. We believe neither of these things of phlogiston, therefore from our point of view the law intended by (*b*) is false.)

Sentences (*c*) and (*d*) are used in the context of an experiment we should describe in terms of pouring H_2SO_4 over zinc, when hydrogen is given off. Following the principles adopted for (*a*) and (*b*), we must say that both (*c*) and (*d*) are false. And yet a perfectly recognizable experiment is being described in which some substance is given off, and as it happens that substance is in our theory (and even in our and Priestley's observation) inflammable. The principle of growth directs us to do justice to the difference between the respective truth statuses of (*a*), (*b*) and (*c*), (*d*). But the principle of no privilege directs us not only to find all four sentences false from the point of view of our theory, but to assume that all the typical sentences of our theory (those presupposing the existence of certain theoretical entities and properties, etc.) are likewise false.

That phlogiston is not an unique sort of case from which to induce is indicated by many other examples from more recent science. For example, Field has discussed the relation between the concept of 'mass' in Newtonian theory ($mass_N$) and the proper mass ($mass_p$) and relativistic mass ($mass_r$) in relativity theory, and Fine has discussed the use of 'electron' during the first century of its life in physics.[12] Field's central point is that $mass_N$ is ambiguous between $mass_p$ and $mass_r$. Some sentences in Newtonian theory containing $mass_N$ are true of both $mass_p$ and $mass_r$ (this includes many 'observation sentences'), some are true of one and false of the other, and some are true of neither. Thus, from the point of view of our theory, $mass_N$, like phlogiston, fails to refer, there never were any $mass_N$'s. Nevertheless there were, from the point of view of our theory, true sentences in Newtonian theory. Field suggests explicating this situation by separating the notion of truth from that of reference, and saying that some sentences in past theory can be said to be true although their terms do not refer, for example (*c*) and (*d*), or the sentence 'mass$_N$'s freely accelerate towards the earth'. Or rather, the terms refer *ambiguously* (phlogiston sometimes but not always to hydrogen, mass$_N$ to mass$_p$ or mass$_r$ in different contexts). We cannot, therefore, take (*c*) and (*d*) as simply true, or be forced to choose in Newtonian theory which uses of mass$_N$ refer to mass$_p$ and which to mass$_r$—mass$_N$ refers in a sense to both and to neither. We must appeal to the translation process, and use the principles of charity and inductive inference to translate Newton's or Priestley's foreign theoretical vocabulary, so that as many as possible of those sentences that seem to be well confirmed by their own observation sentences come out well confirmed, on the same evidence, in our best theories.

It will sometimes happen, of course, that truth relations internal to one theory are not invariant to the translation process. For example, suppose there are two sentences *X*, *Y* containing mass$_N$ in Newtonian theory, which go into true sentences under translation, but where mass$_N$ becomes mass$_p$ in one sentence and mass$_r$ in the other. Then some Newtonian sentences obtained from *X* and *Y* by truth-preserving operations in Newtonian theory, for example substitution of equivalences for mass$_N$ from *X* to *Y*, will not go over into true sentences in the translation. But this seems to be a general consequence of any referential theory of truth that depends, as the present one does, on consensus, because consensus itself is not necessarily truth preserving under equivalence operations even within one natural language, as is evident from the various paradoxes of implication, confirmation, etc.

The translation process can be usefully compared with the so-called 'principle of correspondence' which has been held to govern relations between difference theoretical systems. As sometimes expressed, this principle requires that, when a theory already successful in a given domain is reduced to or superseded by another theory, usually applicable in a wider domain, the second theory must tend in the mathematical limit to become equivalent to the first in the domain where the first is successful. It has been objected that, first, not all theories are in a mathematical form that gives sense to the idea of such a limiting process, and second, that even where they are, it is *concepts* and not only mathematical formulation that change from theory to theory, and the principle as thus expressed has no means of relating the concepts to different theories. As we have seen, it is possible even that entailment relations are not preserved in translation from one theoretical language to another. For example, when special relativity goes over mathematically into Newtonian mechanics for small velocities, the equivalent formulas do not imply equivalent concepts of space, time and mass, and do not show how to relate the quite different conceptual frameworks defining these in the two theories. It is precisely this gap that the principle of charitable translation is needed to fill, and this principle brings the notion of correspondence into a form more closely representative of what scientists generally seem to have understood by 'correspondence' in situations of reduction.

In what sense, then, does science grow? Given the principle of charity in historical interpretation, there is accumulation of true observation sentences in the pragmatic sense that we have better learned to find our way about in the natural environment, and have a greater degree of predictive control over it. This historical induction is not contradicted by the fact that *use* of our new means of control has caused further problems of large-scale systems going out of control: there is only a possibility of nuclear energy going out of control because its component set-ups have come progressively under control. This pragmatic concept of growth can be extended also to some theoretical sentences which are carried over fairly directly from a past theoretical framework to our own, that is, which do not depend for their truth on the existence and classification of particular hypothetical entities, but are nearer to pragmatic predictive test. That the earth is round is true, though it was a highly theoretical inference in the past; that water is composed of discrete molecules of hydrogen and oxygen in definite proportions is true, though we are not able to specify in ultimate terms what exactly molecules and atoms of water, hydrogen, and oxygen *are* (Newtonian, Daltonian, quantum, and relativistic field theories tell different stories about them).

The historical fact that true sentences have accumulated in this pragmatic sense is independent of theoretical revolutions, because the evidence for it depends on what we mean by science: namely the truth in all systems of some observation sentences, and the satisfaction of the principle of correspondence enriched by charitable translation. This formulation of the growth of science does not presuppose privilege for our theory, because it is consistent with replacement of whole conceptual frameworks, including basic classifications and property assignments, while still giving a sense in which some of our theoretical sentences and most of our observation sentences are true. The extrapolation of the historical induction into the future is equally independent of theoretical revolutions, although not of course of the contingent survival of science as we understand it. What we can say is that if our science survives and continues to be subject to theoretical revolution, this is not inconsistent with the continued pragmatic growth of science in the same sense that it has grown in the past.

This is not perhaps the sense in which the lay public, metaphysicians, theologians, and even some theoretical scientists, would like it to grow. For it is not a convergence of ontologies approximating better and better to a description of the true essence of the world, to a final delineation of the pre-Socratic ideal of 'what the world is made of'. It is rather an instrumental growth, as pragmatic as our desire to have true, controllable predictions. Science is after all the Martha, not the Mary, of knowledge.

5. Postscript for historians of science

The consensus theory of truth suggested here has not depended on the assumption of any privilege for the truth of the *theoretical* framework of our science, but rather on the propriety of defining our science in terms of a certain category of *observation sentences* and on a particular *inferential method*. Specifically, it has depended on four assumptions:

(*a*) That there is a set of observation sentences in any natural language which can mostly be taken to be referentially true.

(*b*) That these can mostly be translated from language to

language with the help of charitable historical interpretation, and with preservation of truth.

(*c*) That our science can be specified in terms of a model of inductive inference from observation sentences, and of the overriding goal of cumulative predictive success, which also serve to define what we mean by the truth values of theoretical sentences, and by the growth of scientific knowledge.

(*d*) That the truth of alien theoretical sentences is tested by the charitable translation procedures of (*b*), and by their conformity with our inference patterns and goals.

My claim here is not that an adequate univocal concept of truth depends crucially on acceptance of any particular model of inference, or even of the goal of our science, but rather that *some* account must be given of these if we wish to safeguard our intuitions about the growth of science on the one hand and radical theory change on the other. The philosopher in search of a theory of scientific truth must find a model that is not merely descriptive but prescriptive, or at least explicative (in Carnap's sense) or an ideal type in Weber's sense, since no pure description of what scientists do, or conceive of themselves as doing, is going to yield sufficient consistency for a theory of truth. But at this point the historian of science may well become restive, and it is his doubts that I want to pursue briefly in this section.

First of all, the historian may say, the suggested theory of truth depends heavily on a hermeneutic principle of 'charitable translation', and this is a problematic principle of *historical* method about which more should be said before the theory of truth can be considered complete. Secondly, the historian may complain about the imposition upon past science of the patterns of inference and the goals of our science, arguing that in historical perspective this is just as artificial an imposition as that of the truth of our science, which has been rejected in the principle of no privilege.

To take the second point first, it must be accepted that historical and sociological study alone does not reveal any sharply defined goals and criteria of inference for something called 'science' either in our culture or in cultures of the past.

In any case it would be perverse to maintain that successful predictive control as we conceive it was the implicit goal of early science, whether Greek or even seventeenth-century, when no such systematic control was yet visible or even imaginable. There are of course hints of it, especially in the Baconian tradition, but what is more evident is that the goal of science then was what Habermas calls 'contemplative knowledge',[13] and this has been perpetuated wherever the ideal of correspondence has been adopted as the model of scientific truth. Neither has scientists' understanding of their methods been constant in the past: there have been disputes between inductivists and deductivists, positivists and realists, rationalists and empiricists, even in the same areas of inquiry at the same times and places. As specified by methods and goals, 'science' at best exhibits throughout history a kind of family resemblance continuity, within which privilege cannot be claimed for either one type of method and goal, nor one type of theory.

Once this is accepted, however, it may still be necessary to draw a distinction between the task of the philosopher and the task of the historian. Since it is an intuition of the growth of *our* science that the philosopher is trying to explicate, he is entitled to presuppose our science in a way that may be impermissible for the historian. On the other hand, if we return to the historian's first question, about the nature of charitable translation, things do not look that simple. For the historian also certainly needs such a principle no less than the philosopher, though for different reasons.

Return for a moment to the example of Priestley's phlogiston theory. There has been much discussion centering round the theme, did Priestley discover oxygen? Stripped of its original Whiggish context of the search of priorities in discovery, the question can be interpreted as a request for historical reconstruction of the events as well as the beliefs that led Priestley to claim he had collected what he called 'dephlogisticated air', which allowed flames to burn more brightly, expiring mice to revive, etc. However far the historian tries to reconstruct what happened from Priestley's own works and other contemporary evidence, it would be absurd to suggest that he should ignore what he knows about the chemistry of oxygen and the experimental set-ups that produce it and result from it. Properly used, such knowledge increases his understanding of Priestley and his system. It follows, of course, that historical reconstruction of, for example, alchemical experiments, and also of contemporary rain rituals, and so on, is harder just to the extent that they are further removed from our scientific knowledge, cutting us off from the possibility of understanding to a greater extent than in the case of Priestley's science. This sort of consideration alone would justify some distinction between the history of science and the history of other types of belief system. Here if anywhere in historiography there is need and justification for the hermeneutic principle formulated by Leonard Hodgson in the context of reconstructing the events lying behind the New Testament documents: 'What must the truth be and have been if it appeared like that to men who thought and wrote as they did?'[14] For such a principle to be applicable, the historian needs some justification for taking what we believe about our experience as valid for the reconstruction of the past. In the case of science this requires a theory of truth such as the consensus theory I have outlined.

In conclusion let me make a few more remarks about the task of the historian in the debate about the growth of scientific knowledge. Although I have been a lifelong devotee of the close integration of the study of history and philosophy of science, it does seem to me that the emergence of the hybrid animal 'historian-and-philosopher-of-science' has had some unfortunate consequences, and indicates the need for some revival of a division of labour. Equally unfortunate, and not unrelated to this phenomenon, has been the earlier intellectual and professional distinction of 'historians of science' from historians in general. I have just given an example where I think the philosopher's proper interest in the perennial problem of a concept of truth is distinct from, though complementary to, the historian's interest in particular belief systems and modes of rationality in past science. The historian can aid the philosopher in uncovering alternative concepts of 'science' that have been exemplified in history. But the historian's task can never end there, because history is the study, not of academic disciplines or of belief systems alone, but of men in all the multiplicity of their interactions. It is the increasing realization of this that has pushed some recent history of science towards the study of social, economic and political determinants of science and even of its varied theoretical frameworks. One does not have to subscribe to a Marxist theory of history to find the purely intellectual history of science arid and incomplete. The philosopher has given the historian the concept of the underdetermination of theories by data and inductive inference; it would be odd if the historian saw the resulting problem merely in terms of a search for new types of *rationality* that will restore determination to the history of scientific ideas. An increasing number of excellent studies in the sociopolitical correlations of scientific ideas are showing that such an idealist presupposition is patently untenable.[15] Let historians primarily look after their task of investigating science as part of general history, while philosophers are primarily concerned with analyzing the nature and limitations of our conceptions of truth and rationality. These tasks are not the same.

Notes

1 H. Putnam, 'What is "realism"?' *Proc. Arist. Soc.*, 1975/6, 177.

2 H. Field, 'Tarski's theory of truth', *Journal of Philosophy*, vol. lxix, 1972, 347.

3 As some of the contributors to the discussion at the PSA meeting pointed out, it may clarify things if I accept that the supplement to Tarski's theory required here is a 'verificationist theory of truth'. That general description may be accepted, but it does not of course follow that it will be verificationist in the early positivist sense. A more holistic theory of observation sentences such as those proposed by Duhem and Quine is required. See below, and also my *The Structure of Scientific Inference*, London, 1974, especially chs 1 and 2 (ch. 1 is reprinted as ch. 3 in this volume).

4 W. v. O. Quine, 'Two dogmas of empiricism' in *From a Logical Point of View*, Cambridge, Mass. 1953, p. 20; *Word and Object*, New York, 1960, chs 1 and 2.

5 W. v. O. Quine, 'Natural kinds' in *Ontological Relativity and Other Essays,* New York, 1969, p. 114.

6 For example, *Word and Object*, p. 59, n. 2; D. Davidson, 'Belief and the basis of meaning', *Synthese*, vol xxvii, 1974, 309.

7 I have developed it in my *Structure of Scientific Inference*, and ch. 3 above. There, among other things, I argued that there is no distinction for inductive purposes between the 'theoretical language' and the 'observation language'. So-called 'theretical terms' are not without meaning in the observation language, for if they occur in inductively justified hyptheses they are drawn from the observation language by means of models and analogical argument. This account of theoretical terms is necessary to ensure that theoretical sentences can be given probability values conditionally upon evidence expressed directly in observational terms.

8 I owe the suggestion of these refinements to Richard Sorabji.

9 W. v. O. Quine, *Word and Object*, p. 68, and 'Reply to Chomsky' in *Words and Objections*, ed. D. Davidson and J. Hintikka, Dordrecht, 1969, p. 302.

10 See note 1 above.

11 Details of this case history are most conveniently available in J. B. Conant, 'The overthrow of the phlogiston theory: the chemical revolution of 1775–1789', *Harvard Case Histories in Experimental Science*, vol, i, Cambridge, Mass., 1957, 65.

12 H. Field, 'Theory change and the indeterminacy of reference', *J. Phil.* vol. lxx, 1973, 462; A. Fine, 'How to compare theories, reference and change', *Nous,* vol. ix, 1975, 17.

13 J. Habermas, *Knowledge and Human Interests*, trans. J. J. Shapiro, London, 1972, p. 301 ff.

14 Cf. D. Nineham, *Saint Mark*, London, 1963, p. 52.

15 Cf. ch. 2 above, and among others; P. Forman, 'Weimar culture, causality and quantum theory, 1918–1927: adaptation by German physicists and mathematicians to a hostile intellectual environment,' *Historical Studies in the Physical Sciences*, vol. iii, 1971, 1; W. Coleman, 'Bateson and chromosomes: conservative thought in science', *Centaurus*, vol. xv, 1970, p. 228; D. Holloway, 'Innovation in science—the case of cybernetics in the Soviet Union', *Science Studies*, vol. iv, 1974, 299; B. Barnes, *Scientific Knowledge and Sociological Theory*, London, 1974; D. A. MacKenzie and S. B. Barnes, 'Biometrician versus Mendelian: a controversy and its explanation', *Kolner Zeits fur Soziologie und Sozialpsychologie*, 1975; R. M. Young, 'Malthus and the evolutionists: the common context of biological and social theory', *Past and Present*, vol. xliii, 1969, 109, and 'The historiographic and ideological contexts of the nineteenth century debate on man's place in nature' in *Changing Perspectives in the History of Science*, ed. M. Teich and R. M. Young, London, 1973, p. 344.

III PRAGMATIC AND EVALUATIVE KNOWLEDGE

7 *In Defence of Objectivity*

I

Various intellectual and moral tendencies are currently combining to dethrone natural science from the sovereignty of reason, knowledge, and truth which it has enjoyed since the seventeenth century. Far from being the paradigm of objective truth and control which will make us free of all natural ills and constraints, science is increasingly accused of being a onesided development of reason, yielding not truth but a succession of mutually incommensurable and historically relative paradigms, and not freedom, but enslavement to its own technology and the consequent modes of social organization generated by technology. It is with the intellectual, rather than the moral or practical, sources of these criticisms that I shall be concerned here. I want to try to discriminate among various aspects of the implied attack on scientific objectivity, and to consider how far and in what sense claims to objectivity can be maintained.

During the last half-century much of professional Anglo-American philosophy of science has been devoted to detailed development of the internal logic of natural science based on empiricist criteria, and also to attempts to show how this logic applies also in the social sciences and in the study of history. Suggestions such as those deriving from the traditions of Dilthey or Weber to the effect that there are other modes of knowledge than the empiricist were sometimes actively resisted but more usually totally disregarded. The corollary was that *if* the human sciences are to attain knowledge-status at all, *then* their method must conform to some acceptable modification of that of the natural sciences, whose own method, it was claimed, was in all essentials thoroughly understood. During the same period continental philosophy has on the whole ignored these technical analyses of science. Sometimes, as in Husserl and Heidegger, natural science was the subject of negative assessment of its credentials and value as a claim to knowledge; more usually a late nineteenth-century form of instrumentalism has been uncritically accepted as the last word about such claims. However, in the postwar period two continental traditions have become more self-conscious about problems of epistemology and method, although neither of them has been primarily concerned with natural science. These are the mainly Protestant schools of biblical exegesis, and the Marxist-oriented schools of political and social philosophy. In both traditions the term 'hermeneutic' has been adopted to indicate concern for knowledge as *interpretation*, sometimes explicitly distinguished from what is taken to be the direct, literal, uninterpreted modes of description proper to the natural sciences.

The basic problem of hermeneutics may be briefly expressed by an analogy more familiar on the English philosophical scene, namely the so-called 'paradox of analysis'. Just as a paradox seems to arise when more precise logical or conceptual tools are used to analyze ordinary vague usage of language, because the product of such analysis is not then identical with what was analyzed, so in a much more general sense a 'hermeneutic circle' arises when the language, categories and frameworks of our own culture are used to interpret and understand alien texts, alien cultures, and even other individuals and groups in our own culture or society. This is because the language and thought forms we are studying are not in themselves intelligible without interpretation, but our own language and thought forms are not adapted to fit them, therefore interpretation is always problematic and accompanied by distortion. The hermeneutic circle is held to arise particularly in studies of the human rather than the natural world, just because, it is claimed, human subjects have their own understanding and interpretation of their states and activities, whereas physical and biological nature does not. Nature can therefore be understood externally and objectively in terms of our categories without distortion, human societies cannot.

Apart from such characterizations as this, there is as yet little detailed investigation of the credentials of the hermeneutic method, certainly not such as would satisfy Anglo-

American trained philosophical analysts. There is, however, an impressive corpus of examples of the problems to which it is claimed to be relevant, ranging through interpretations of New Testament and other esoteric texts, studies of primitive ritual and myth, and in general cross-cultural and cross-ideological investigations, to the historical and contemporary study of psychiatry and the modes of madness. This is not the place, neither do I have the capacity, to attempt a detailed analysis of hermeneutic methodology. What I want to do is rather to compare its implied distinction between methods in the natural and human sciences with a potentially more radical development within the historical and philosophical analysis of natural science itself. For the imperialism previously claimed for natural science in the empiricist tradition has now turned in some quarters into its opposite, namely an assimilation of natural science itself to something approaching the hermeneutic critique. This critique comes both from philosophers of science dissatisfied with logical empiricist accounts of the structure of science and from historians of science who have been brought to question the theory of a 'demarcation' of science from other attitudes to and theories of the natural world, in the light of the similarities and continuities between 'science' and 'pre-science' or 'non-science' that can be found in its history. Study of witchcraft cults among the Azande is not apparently so different in its methodology and philosophical moral from, say, study of Stoic physics.

It is covenient to take as starting-point a perceptive discussion by Jurgen Habermas of the similarities and differences between empirical and hermeneutic method in his book published in English as *Knowledge and Human Interests*.[1] I shall consider first a group of distinctions concerning traditional problems of the language and epistemology of science, taken from his exposition of Wilhelm Dilthey. These are distinctions that I believe are made largely untenable by recent more accurate analyses of natural science. They may be briefly summarized in the following five points. (In considering these points in relation to hermeneutic method, it helps to keep in mind the least controversial type of application of that method, namely the study of history consider some standard problem of interpretation, for example, the causes of the First Crusade.)

1. In natural science experience is taken to be objective, testable, and independent of theoretical explanation. In human science data are not detachable from theory, for what count as data are determined in the light of some theoretical interpretation, and the facts themselves have to be reconstructed in the the light of interpretation.
2. In natural science theories are artificial constructions or models, yielding explanation in the sense of a logic of hypothetico-deduction: *if* external nature were of such a kind, *then* data and experience would be as we find them. In human science theories are mimetic reconstructions of the facts themselves, and the criterion of a good theory is understanding of meanings and intentions rather than deductive explanation.
3. In natural science the lawlike relations asserted of experience are external, both to the objects connected and to the investigator, since they are merely correlational. In human science the relations asserted are internal, both because the objects studied are essentially constituted by their interrelations with one another, and also because the relations are mental, in the sense of being created by human categories of understanding recognized (or imposed?) by the investigator.
4. The language of natural science is exact, formalizable, and literal; therefore meanings are univocal, and a problem of meaning arises only in the application of universal categories to particulars. The language of human science is irreducibly equivocal and continually adapts itself to particulars.
5. Meanings in natural science are separate from facts. Meanings in human science are what constitute facts, for data consist of documents, inscriptions, intentional behaviour, social rules, human artefacts, and the like, and these are inseparable from their meanings for agents.

It follows, so it is held, that in natural science a oneway logic and method of interpretation is appropriate, since theory is dependent on self-subsistent facts, and testable by them. In human science, on the other hand, the 'logic' of interpretation is irreducibly circular: part cannot be understood without whole, which itself depends on the relation of its parts; data and concepts cannot be understood without theory and context, which themselves depend on relations of data and concepts.

There are obscurities in the way these points have been set out which badly need investigation, particularly in relation to the concepts of 'interpretation' and 'meaning'. It is immediately apparent, for instance, that there is an ambiguity in the way 'meaning' has been used in relation to natural and human science respectively. 'Meaning' in natural science presupposes an account of the empirical reference of terms and of their intensional connotations within a scientific theory. The concept of 'meaning' in the hermeneutic sciences, on the other hand, is much richer, for it carries implications for the data that go beyond an external semantics of language. Data in the human sciences are said to be themselves constituted by 'meanings' in virtue of being the products of human language and intentions. Again, it is implied in the contrast drawn between the natural and human sciences that there is an unproblematic sense in which insight can be gained into human intentions, rules, and meanings which is different from the purely external understanding of nature. But it is by no means clear that this sense is so unproblematic. The thought forms of alien cultures may be so foreign to our own that it might make sense to say that I understand my dog, or even my chrysanthemums, better than I understand those people. This is not to say, of course, that I fully know what it is to understand my dog, if by this is meant more than an ability to teach him tricks and to predict his external behaviour. But it does suggest that the notion of understanding 'meanings' in some of the alleged applications of hermeneutic method need much more investigation. It is precisely one of the dilemmas facing students of alien thought and culture that the distinctions between external behaviour and meaning, cause and reason, are far from easy to draw.

Let us, however, concentrate for the moment on the natural science half of the dichotomy. What is immediately striking about it to readers versed in recent literature in philosophy of science is that almost every point made about the human sciences has recently been made about the natural sciences, and that the five points made about the natural sciences presuppose a traditional empiricist view of natural science that is almost universally discredited. In this traditional view it is assumed that the sole basis of scientific knowledge is the *given* in experience, that descriptions of this given are available in a theory-independent and stable language, whether of sense data or of common sense observation, that theories make no ontological claims about the real world except in so far as they are reducible to observables, and that causality is reducible to mere external correlations of observables. It is no novelty that all these empiricist theses have been subject to much philosophic controversy. It has been accepted since Kant that experience is partly constituted by theoretical categories, and more recently than Kant it has been generally held that these categories are not *a priori*, but are conjectured by creative imagination, having a mental source different from experiential stimuli. Moreover, the work of Wittgenstein, Quine, Kuhn, Feyerabend, and others has in various ways made it increasingly apparent that the descriptive language of observables is 'theory-laden', that is to say, in every empirical assertion that can be used as a starting-point of scientific investigation and theory, we employ concepts that *interpret* the data in terms of some general view of the world or other, and this is true however apparently rooted in 'ordinary language' the concepts are. There are no stable observational descriptions, whether of sense data, or protocol sentences, or 'ordinary language', in which the empirical reference of science can be directly captured. Paralleling the five points of the dichotomy, we can summarize this post-empiricist account of natural science as follows:

1. In natural science data is not detachable from theory, for what count as data are determined in the light of some theoretical interpretation, and the facts themselves have to be reconstructed in the light of interpretation.
2. In natural science theories are not models externally compared to nature in a hypothetico-deductive schema, they are the way the facts themselves are seen.

3. In natural science the lawlike relations asserted of experience are internal, because what count as facts are constituted by what the theory says about their interrelations with one another.
4. The language of natural science is irreducibly metaphorical and inexact, and formalizable only at the cost of distortion of the historical dynamics of scientific development and of the imaginative constructions in terms of which nature is interpreted by science.
5. Meanings in natural science are determined by theory; they are understood by theoretical coherence rather than by correspondence with facts.

It follows, so it is held, that the logic of science is necessarily circular: data are interpreted and sometimes corrected by coherence with theory, and, at least in less extreme versions of the account, theory is also somehow constrained by empirical data. The resemblances between this account and the hermeneutic analysis of the human sciences seems so close that, among the more extreme post-empiricists, Feyerabend at least has drawn the explicit conclusion that scientific theories and arguments are closely analogous to the circular reinforcement of beliefs, doctrines, documents, and conditioned experience that may be found in some religious groups, and in political party lines and their associated techniques of propaganda.[2]

II

There are some features of this post-empiricist analysis which I do not want to dispute here. I take it that it has been sufficiently demonstrated that data are not detachable from theory, and that their expression is permeated by theoretical categories; that the language of theoretical science is irreducibly metaphorical and unformalizable; and that the logic of science is circular interpretation, reinterpretation, and self-correction of data in terms of theory, theory in terms of data. Such a view of science is by no means new: it is to be found in all essentials in those fathers of inductive science, Francis Bacon and Isaac Newton. I shall later suggest a model of natural science as a *learning device* that can be made to represent such an account without abandoning the essentials of empiricism, and which shows that the logic of science implied in the account is virtuously rather than viciously circular.

There is, however, a further aspect of both the empiricist and post-empiricist accounts of natural science that has not yet been touched on, and which is of crucial importance for the comparison of natural and human science. This is the question of scientific *truth*, and the consequent credentials of natural science as a form of objective knowledge. In the early period of modern science it was plausible to believe, and indeed it was believed by both Bacon and Descartes, that natural science would be a continuously progressive, cumulative, and convergent approach to truth, where truth was understood as correspondence between a system of objective knowledge and the real world. It was therefore reasonable to adopt a *realist* interpretation of scientific theory as that which progressively discovers or uncovers the hidden essences of nature. It soon became apparent in the subsequent history of science, however, that there is no such cumulative approach to description of a real world of essences by scientific theory. The conceptual foundations and premisses of theories undergo continuous and sometimes revolutionary change, and this occurs not merely *before* the so-called scientific revolution in method of the seventeenth century, but subsequently, when the method of science remained comparatively stable. The succession of theories of the atom, and hence of the fundamental nature of matter, for example, exhibits no convergence, but oscillates between continuity and discontinuity, field conceptions and particle conceptions, and even speculatively among different topologies of space.

The empiricist response to this instability of theory has been the positivist or *instrumentalist* view of science as constituted essentially by accumulating knowledge of phenomena or observables, rather than of the fundamental but hidden nature of things. This is the kind of knowledge that issues in technical application, the cumulative character of which cannot be in doubt. Thus the claim of science to yield objective knowledge comes to be identified with the cumulative possibilities of instrumental control rather than with theoretical discovery, and this in fact is the conclusion drawn by Habermas and most other hermeneutic philosophers when they come to compare the forms of objectivity of the natural and human sciences. However, this conclusion of empiricism has also come under fire from the post-empiricists, whose reinterpretation of the role of scientific theory also reopens the old debate between realism and instrumentalism.

Two features of the new analysis are relevant to this debate. First, it is held that successive theories so permeate observation statements that there is no stable observation language in which the empirical reference of science can be directly captured. It follows on this view that the objective corpus scientific knowledge pointed to by instrumentalism does not form a neutral and accumulating expression of 'facts' discovered by science. Instrumentalism can no longer interpret the truth claims of science as a body of empirical statements, but can at best point to the pragmatic effects of science to indicate its form of empirical objectivity. On the other hand, talk of the 'truth' of science, and of the ontology of objects which it presupposes, becomes wholly internal to scientific theory itself. Truth and existence claims are determined, not by the world, but by the postulates of theory: for our physics *there are* fundamental particles and fields, a space-time continuum, forces and persisting physical objects; for other cultures *there are* spirits, witches, telepathic communications, persons not uniquely and continuously space–time locatable, and so on and so on. It has been held to follow in this view that the currently accepted theory must supersede in all its implications even a natural descriptive language that was pervaded by a previous theory. For example, the assertion 'the table is hard and solid' must be held to be *false* relative to the new language developed by physics, because current physical theory asserts that the table is a field of elastic repulsive forces, and is mostly empty space. Sometimes the corollary is also explicitly adopted, namely that the 'currently accepted theory', which thus determines the categories of observation, is accepted on wholly non-empirical grounds, and is in fact indistinguishable from myth or metaphysics. There is no room in this view for an objective account of scientific knowledge in terms of accumulations of true empirical statements, either theoretical or observational.

A more conservative conclusion from post-empiricist premisses is that not current theory, but current 'common sense' observation sentences, should be given privileged status. In the light of critical demolition of the notion of theory-independent observation sentences, this view will not now be held on grounds of the relative stability of the observation language, but of the demonstrable *in*stability of theories. If every theory is destined to prove inadequate and to be replaced by a theory differing radically in its concepts and laws, then, it may be argued, we are likely to have more direct evidence for, and to be more convinced of the truth of, common sense descriptions than any theoretical descriptions. This is the view not only of instrumentalists in the philosophy of science, but also of all 'ordinary language' analysts who resist the claim that scientific theory may *change* 'what it is correct to say' in ordinary language, and of all phenomenologists who hold that some phenomenological reduction of immediate human experience is more fundamental than the 'objectifications' of science. Ironically enough, this is a view that also in its way implies a relativity of science to theory. For as soon as it is admitted, as it must be in the light of the findings of history of ideas and of anthropology, that conceptually very different 'common sense' languages may be viable, and that a given language may radically change, the language appealed to by the 'common sense' school must be conceived to change *irrationally* with external circumstances, and not as a result of any discovery or rational consideration of empirical truth yielded by science. This second view leaves no room for accumulating objective description of the empirical either.

There is, however, a third possibility, which does more justice to the sequence of theory systems as we actually find them in the history of science. This is the view that successive theories supersede and reinterpret their predecessors, but without rejecting the empirical discoveries that they embody. The table can still be said in some sense to be solid, and this assertion retains some of the implications it previously had: balls will bounce on it, heads will crack on it. But other implications of the previous matter theory are now false: for example that it, or any part of it, is indefinitely divisible into homogeneous pieces of stuff, that it has mathematically sharp surfaces or edges, and so on. Moreover, the new theory does not just contradict parts of the old theory, it also explains why the old theory was as good as it was and what its limitations are: that it is a good approximation only in the case of macroscopic objects, moderate velocities, etc. This implies that something remains constant from theory to theory. What that something is can best be expressed by pointing to classifications of what count as similar systems subject to the same laws, and the forms of those laws or approximations to them. For example, that the planets, the earth, and stones falling on the earth, are similar types of body and satisfy the same laws, was a discovery made in the seventeenth century which has been maintained through the revolution of modern physics, and so have the approximate forms of these laws within certain empirical limits. Such discoveries have not been affected by subsequent radical conceptual changes in the theory of space and time, or in the understanding of mass and its natural motions, which affect all these bodies alike. Lawlike structures and similarities of nature between physical systems have been maintained and are cumulative. Theoretical interpretations of what the natures of these systems absolutely are, are not. Hence even on such a moderate interpretation of post-empiricism, science must still be said to yield phenomenal or instrumental rather than theoretical knowledge.

III

Post-empiricist analyses of science have placed more emphasis on theories than their empiricist predecessors, but in the end they support rather than undermine the conclusion that natural science is essentially instrumentalist. On the relative value to be given to science as aiming at explanatory theories, and science as the basis for instrumental knowledge, however, Habermas parts company both with Husserl and Heidegger, and with some of the post-empiricists, notably

Feyerabend. On the one hand, Feyerabend regards a proliferation of competing imaginative theories as the mainspring of scientific activity, while reducing pragmatic application to a trivial byproduct of this development. In his view, claims for the external truth or objectivity of scientific theory are damaging, since they easily degenerate into dogmatism by circular reinforcement of theory by experience conditioned by theory. Habermas, on the other hand, while agreeing that theory has no claim to objectivity as such, nevertheless maintains the more conservative view that it is just the possibility of technical exploitation that guarantees the value and objectivity of natural science.

It is indeed a main motivation of Habermas' argument to direct attention to the *human interests* served by natural and human science respectively, and to their respective criteria of success and failure, or, as he puts it, to their respective forms of objectivity. In natural science the interest is in exploitable technical control, and the character of natural science as 'objective', 'detached', and 'value-free' is itself a value characteristic derived from the human decision to develop a form of knowledge which is thus technically exploitable. The sanction of failure is unsuccessful feedback from active prediction and test. Successful feedback depends on the presupposition that the conditions of human nature and its environment remain sufficiently the same: the natural sciences 'grasp reality with regard to technical control that, under specified conditions, is possible everywhere and at all times'.[3] Thus Habermas rejects Marcuse's claim that a new form of society would entail a new science which 'would arrive at essentially different concepts of nature and establish essentially different facts'. On the contrary Habermas believes only that a new *attitude* to science is possible:

> The idea of a New Science will not stand up to logical scrutiny any more than that of a New Technology, if indeed science is to retain the meaning of modern science inherently oriented to possible technical control. For this function, as for scientific technical progress in general, there is no more 'humane' substitute.[4]

In this defence of the objectivity of natural science as technical control, Habermas again rejects the claim that scientific *theory* can describe objective natural reality in favour of an instrumental objectivity guaranteed by control. Marcuse may well be correct in holding that a revolutionary society would generate a new conceptual view of nature, as indeed has happened in English society, for example, in the Renaissance, Restoration, Enlightenment, and Industrial periods. However Habermas' point seems to be that whatever theoretical system is adopted, there will be similar perennial and universal possibilities of instrumental control, and moreover, he holds that a theory of nature going beyond that technical interest to masquerade as a 'pure' ontology is an illusion—possibly a dangerous illusion, since it seems to provide the ideological justification for unbridled engineering both natural and social. In his rejection of realistic interpretations of science as dogmatism or ideology, Habermas is at one with Feyerabend, but it is easy to imagine how Habermas would respond to Feyerabend's rejection of 'mere' technology as an essential ingredient of science. The technologically unconstrained proliferation of theories and ontologies of the natural world recommended by Feyerabend would be ideological opium for the masses alienated and bored by pervasive technology: circuses without even the corresponding bread.

In Habermas' interpretation the forms of objectivity of natural and human science are not transcendental, but are dependent on the value or interest put upon their respective activities by a human community. Whereas the interest of natural science is technical control, requiring skills in the interrogation of nature, the interest of human science is social consensus, mutual communication and practical effectiveness in social organization, and this requires skills of personal understanding. The guarantee of objectivity in human science is the participation in *dialogue* between investigator and investigated, in which *reciprocal* interaction occurs. The sanction of failure is disturbance of consensus and breakdown of communication. It is clear that the consensus referred to is not the forced consensus of the totalitarian state, since this precludes communication and reciprocal influence. It is rather the consensus produced by partners in dialogue, both of whom may be freely persuaded and changed by the encounter. Neither is it Dilthey's concept of empathy or *verstehen*, in which the investigator claims to enter the mind of his subject and think his thoughts after him, for this presupposes that the investigator's own world (out of which he has artificially abstracted himself) does not impinge on and remains unchanged by the encounter.

The model of dialogue as a form of objectivity is unfamiliar and somewhat shocking to those accustomed to empiricist presuppositions, but it is one of the few viable alternatives to the model of natural science in dealing with the human sciences. An illustration from the historiography of science, which is itself a human science, may indicate how it helps to illuminate certain problems of interpretation. I take an example, which I have developed elsewhere, from a recent debate about the received tradition of historiography of sixteenth- and seventeeth-century science. In an article entitled 'The hermetic tradition in Renaissance science', Frances Yates has expressed an entirely proper desire not to interpret the science of the past 'from the solely forward-looking point of view . . . misinterpreting the old thinkers by picking out from the context of their thought as a whole only what seems to point in the direction of modern developments'.[5] Miss Yates asks for a proper balance between this point of view and a study which takes more account of the historical context of ideas at the time. In a relativist climate it is easy to distort such a balanced approach into a refusal to evaluate the science of the past at all in relation to what is now believed to be true, or to discriminate rationality and empiricism in past thought from such philosophies of nature as hermeticism, alchemy, numerology and magic. Sceptical conclusions regarding the 'objective' character of scientific knowledge have been held to follow. However, according to the model of historiography as dialogue, such conclusions are illicit. For a historian operating according to this model, neither the anachronistic reconstruction of past science in the light of modern theories and modern evidence, nor the deliberate suppression of these in the attempt to become a 'seventeenth-century man', are satisfactory or indeed possible. What is required is a sympathetic attempt to enter into seventeenth-century thought forms and problems *without* abandonment of the criteria provided by subsequent developments.

History of science, like all history, is in principle written anew in every generation. Historical interpretations are irreducibly relative to the historian and his time, but it does not follow that they are relativ*ist*, if by this is meant that there are *no* external criteria for the evaluation of past science. On the contrary, there are our criteria as they have emerged in the course of history. In our study of the science of the past we may not irresponsibly neglect them, for they constitute our side of an objective dialogue.

Whether this model of dialogue turns out under more detailed investigation to be entirely successful or not, the attempt to spell out a methodology of human science shows at least two things. It shows that any assimilation of the methodology of natural to that of human science does not entail that both methodologies are non-objective, since the task of a hermeneutic analysis is precisely to make explicit the conditions of objectivity of the method of dialogue. My example of the interpretation of Renaissance science is itself a brief attempt at just such a hermeneutic analysis. On the other hand, the dialogue model also suggests that complete assimilation of the two kinds of methodology will fail, because nature cannot be regarded as a partner in dialogue. An oversimple dichotomy between natural science, on the one hand, and the objectivity of understanding-in-dialogue on the other is reminiscent of Collingwood's conclusion that non-human subject matters are not genuine subjects of knowledge or understanding, because not capable of participating in dialogue. This is not Habermas' view, since he places high objective value on technical control, nevertheless it is a tempting interpretation of his view, because in the end he fails to carry through in detail an analysis of what is involved in technical control and to examine what its limitations are. In the concluding part of this chapter I shall raise some questions about this instrumental model of natural science, and suggest that the relation between it and the hermeneutic model is not so much a dichotomy as a continuum.

IV

In discussing natural science Habermas makes frequent use of the concepts of successful prediction, feedback, and self-correction. In effect this is to appeal to a model of natural science as a *learning machine*.[6] It is not difficult to incorporate most of the features of natural science as at present understood into such a model. The presence of feedback loops in a learning machine allows for the circular self-correction of theory by experience and experience by theory that is demanded by interpretation of science as theory-laden. 'Experience' must be regarded in the model as the input or physical stimuli impinging upon the machine from its environment. The process of describing experience in intersubjective language by the scientific community is representable as the coding of the input into machine language according to whatever categories have been programmed into the machine from the current natural language. Doubtless the coding devices will *also* be subject to modification in the light of feedback from successful and unsuccessful learning by the machine of its environment, just as the natural descriptive language of a human learner may be so modified. Thus the physical stimuli themselves need not be directly expressible in any stable language, and it must be a *hypothesis* that they themselves remain sufficiently stable for what is learned by the machine to be applicable and testable on future occasions. In the case of a learning machine in which we can investigate both the mechanism and its environment, we know what some of the conditions of successful learning are. There must be sufficient possibility of detailed *test* to reinforce correct learning; the environment must be sufficiently stable for the self-corrective learning process to *converge*; and there must not be such strong action by the machine on its environment that either it exhibits no convergence, or what it learns is just an artefact of the machine itself. Without such constraints on the environment, feedback mechanisms are liable to go into unstable oscillation. Habermas' objectivity of technical control presupposes that in the subject-matter of natural science, these conditions are satisfied. The last condition is clearly not satisfied in those sciences he describes as hermeneutic, since these are precisely characterized by strong reciprocal interaction between investigator and investigated, or in terms of the model, between machine and environment. In these sciences also the possibility of detailed test and a sufficient stability of environment will sometimes not be present either.

It would be misleading, however, to conclude that the model of learning is quite irrelevant to the human sciences. In the first place, the human sciences are bound to use some of the techniques developed by the natural sciences, and have as good a claim to objectivity in these respects as any natural science. Dating of archeological findings, and of manuscripts, and reconstruction of historical events from circumstantial evidence are obvious examples. Secondly, in describing a learning machine, nothing need be said about the character of the empirical input, except that it is presumed that assertions are made about it in an intersubjective language. But this does not restrict expressions of the input to phenomenalist protocol sentences nor to positivist observation statements. They may, if intersubjectively acceptable to the scientific community, also include sentences ascribing intentions, motives, and emotions to human beings, for these are commonly used descriptively of overt behaviour, and they are subject to test and correction by well-known processes of ordinary observation. Again, the model of the learning machine is flexible enough to take account of some of the 'subjective' elements in both natural and human science, by the device of self-corrective feedback loops. There are cases in the human sciences, just as there are in the natural sciences, where apparent strong interference by the investigator on his subject-matter may itself be allowed for and corrected if a sufficiently comprehensive theory of the relevant processes is available. In arguing for the unpredictable effects of interaction with the subject-matter, hermeneutic philosophers often compare the situation in the human sciences with the uncertainty principle in quantum physics, where the attempt to measure the position of a fundamental particle is said to interfere irreducibly with its momentum, and vice versa. But the analogy is not an apt one, for our information about this kind of interference comes not from direct observation, but from a complex theory of fundamental particles, other aspects of which *are* known by the usual objective learning process. Similarly, although the logical possibility of irreducible interference can be understood in terms of the learning model, it is not enough in itself to prove that particular parts of the human sciences are opaque to the mode of objectivity appropriate to the natural sciences. It is true that the calculations of the learner may be upset by the presence of the anthropologist in the tribe, or the educational theorist in the school, or the TV camera at the civil disturbance, but on the other hand it may sometimes be the case that such interaction can be minimized and allowed for. The possibility of degrees of independence and objectivity should be recognized, and it is the task of a philosopher of the human sciences to spell these out in detail in particular cases.

The conditions of learning and control, then, are sometimes satisfied in the human sciences and sometimes they are not. Conversely, it may be asked whether they are universally satisfied in the natural sciences. Certainly instrumentalists are right in concluding that they cannot be used to guarantee the objectivity of theoretical science, for we have already seen that it is difficult to make sense of a claim that scientific theory yields objective empirical knowledge unless the succession of theories can be said to be cumulative. Theories are neither cumulative in fact, nor does it seem that such accumulation is a necessary condition for science to be a learning process. If the aim of science is essentially to enable man to learn his way about in his environment, then the only necessary condition for its success is efficiency of learning. As far as we can tell, the learning machine that has been described will continue to learn in a certain kind of stable environment. But we have no idea what is the most efficient method of learning even in such an environment, for the problem of finding theories in terms of which we can learn never has a unique solution. It may be that the quickest learning will take place by frequent and radical changes of theories, or that insertion of some randomness into the machine's selection of best theories and predictions may be advantageous.

Moreover, violation of the conditions of learning themselves is not confined to the human sciences, for the possibility of detailed test, the stability of the environment, and the absence of interaction between machine and environment are not guaranteed by the fact that the subject-matter of natural science is non-human. There are many reasons, ranging from the practical impossibility of detailed test over sufficiently large regions of space and time, to social and moral restraints upon experimentation with the natural environment, which may inhibit efficient working of science conceived as a learning machine. Cosmology and biology cannot be excluded from the domain of natural science, and yet they only imperfectly satisfy the conditions of learning and control. We are left with a problem about the form of objectivity of large areas of natural science that seem to evade both the analysis in terms of learning, and the hermeneutic model of personal dialogue.

It is possible of course that it must just be accepted with natural piety that there is no form of objectivity appropriate to theoretical science. However, since at least a beginning has been made towards the analysis of an objective hermeneutic method appropriate to the human sciences, and since there are at least some features of the natural sciences that exhibit some features of that method, it is permissible to hope that dichotomy is not the last word. In conclusion I shall briefly suggest two reasons why hermeneutics may yet prove to be more important for natural science than has so far been apparent.

First, the view of nature as merely behaviourally known, and of man as internally known, implies a separation of man from nature which is itself an ontological belief. It is indeed the converse of that type of naturalism which has sought to totally assimilate man to nature, and which has claimed, no doubt illicitly, the support of natural science itself. But neither naturalism nor its converse seem to be justified as a consequence of *natural* science. Justification of either view would have to be sought in terms of a method adequate also for the human sciences, and if the dialogue model is taken as that method, it might at least suggest that the understanding of man implies an understanding of related biological nature, and conversely. It is impossible in studying theories of evolution, ecology, or genetics, to separate a mode of knowledge relating to technical control from a mode relating to the self-understanding of man. This is not just to assert that human values will be involved in *applications* of these theories, though that is true too; it is also, and more centrally for the present discussion, to assert that the very categories of these theories, such as functionality, selection, survival, are infected by man's view of himself.

Secondly, as is suggested by these examples, and has been abundantly demonstrated in the history of all natural sciences, theories have always been expressive of the myth or metaphysics of a society, and have therefore been part of the internal communication system of that society. Society interprets itself to itself partly by means of its view of nature. Even to deny the propriety or relevance of this is to hold a view of man's relation to nature, namely their total separability. This is a sense in which nature does indeed partake in the dialogue of man with man, and can itself be said to be informed by human meanings and subject in its theoretical aspects to hermeneutic methodology.

Notes

1 J. Habermas, *Knowledge and Human Interests*, trans. J. J. Shapiro, London, 1972. An excellent analysis by an English-speaking philosopher of similar themes is to be found in Charles Taylor's 'Interpretation and the sciences of man', *Rev. Met.* vol. xxv, 1971, 3.

2 See in particular P. K. Feyerabend, 'Classical empiricism', in *The Methodological Heritage of Newton*, ed. R. E. Butts and J. W. Davis, Oxford, 1970, p. 150; 'In defence of classical physics', *Studies in the History and Philosophy of Science*, vol. i, 1970, 59; and 'Against method', in *Minnesota Studies in the Philosophy of Science*, vol. iv, ed. M. Radner and S. Winokur, Minneapolis, 1970, p. 17.

3 *Knowledge and Human Interests*, p. 195.

4 J. Habermas, 'Technology and science as "ideology"' in *Towards a Rational Society*, trans. J. J. Shapiro, London, 1971, pp. 86, 88.

5 *Art, Science, and History in the Renaissance*, ed, C. S. Singleton, Baltimore, 1968, p. 270. I have discussed this example in 'Hermeticism and historiography' in *Minnesota Studies in the Philosophy of Science*, vol. v, ed. R. Stuewer, Minneapolis, 1970, p. 134, and in ch. 1 above.

6 I have developed this learning model in more detail in 'Duhem, Quine, and a new empiricism' in *Knowledge and Necessity*, Royal Institute of Philosophy Lectures, vol. iii, London, 1970, p. 191, and in ch. 5 above.

8 *Theory and Value in the Social Sciences*[1]

I

Many reasons have been given for supposing that the social sciences require different kinds of method and justification from the natural sciences, and conversely for supposing that these methods and justifications are or ought to be the same. I don't want to rehearse all these arguments here, but rather to concentrate on two features of the *natural* sciences which already suggest that the conventional arguments about similarities and differences are inadequate. These features can be roughly summed up in the by now fairly uncontroversial proposition that all scientific theories are *underdetermined* by facts, and the much more problematic propositions that, this being the case, there are further criteria for scientific theories that have to be rationally discussed, and that these may include considerations of value.

Whether the natural and the social sciences are seen as similar or different depends of course on the view we take of the natural sciences. The view I am going to presuppose, but not argue here, is that made familiar in recent post-deductivist discussions, with the addition of a crucial pragmatic dimension.[2] Let me summarize as follows:

1. Theories are logically constrained by facts, but are underdetermined by them: that is, while, to be acceptable, theories should be more or less plausibly coherent with facts, they can be neither conclusively refuted nor uniquely derived from statements of fact alone, and hence no theory in a given domain is uniquely acceptable.
2. Theories are subject to revolutionary change, and this involves even the language presupposed in 'statements of fact', which are irreducibly theory-laden: i.e., they presuppose concepts whose meaning is at least partly given by the context of theory.
3. There are further determining criteria for theories which attain the status of rational postulates or conven-

tions or heuristic devices at different historical periods—these include general metaphysical and material assumptions, for example, about substance and causality, atoms or mechanisms, and formal judgments of simplicity, probability, analogy, etc.

4. In the history of natural science, these further criteria have sometimes included what are appropriately called value judgments, but these have tended to be filtered out as theories developed.

5. The 'filtering-out' mechanism has been powered by universal adoption of one overriding value for natural science, namely the criterion of increasingly successful prediction and control of the environment. In what follows I shall call this the *pragmatic criterion*.

Points 4 and 5 need further explanation.

Value judgments related to science may be broadly of two kinds. They may be evaluations of the *uses* to which scientific results are put, such as the value of cancer research, or the disvalue of the nuclear bomb. But they may also be evaluations that enter more intimately into theory-construction as *assertions* that it is desirable that the universe be of such and such a kind *and* that it is or is not broadly as it is desired to be. Examples of positive evaluations of what is the case are: belief in the perfection of spherical symmetry, and consequent belief that the heavens are spherically symmetrical; belief that men ought to be and therefore are at the physical centre of the universe, and that they are biologically superior and unique among organisms; belief that mind is devalued by regarding it as a natural mechanism, and therefore that mind is in fact irreducible to matter. An example of negative evaluation of what is the case is the Marxist belief that in this pre-revolutionary stage of the class struggle various elements of social life that look like valuable supports of social stability are to be unmasked as in fact being obstacles to the desirable revolution. In the light of such a belief, for example, the immiseration of the proletariat becomes a positive value, and tends to become the essential category in terms of which complex social facts are described.

It is the second type of evaluation in science that I shall be concerned with. All examples of this type issue in assertions rather than imperatives, and hence involve a transition from *ought* judgments to *is* judgments. Are they not therefore immediately condemned as illicit in any form of scientific argument? In reply to this objection two points can be made. First, there is no doubt that there are historical examples in which the genetic fallacy was not seen as a fallacy, so that in describing the thought processes involved in such examples the historian at least has to recognize forms of quasi-inference such as those just sketched. But the second and more important consideration depends on point 3 above, namely that since there is never *demonstrative* reasoning from evidence to theory, further determining criteria may well include factual judgments about the way the world is, and these are sometimes based persuasively on judgments of how it ought to be. There is no fallacy of logical inference, for logical inference is not appropriate here; there is rather the choice of some hypotheses for consideration among many other possible ones, in the hope that the world will be found to be good as the accepted value system describes the good.

In the case of the natural sciences, however, it may well be objected that the evaluative and teleological beliefs of past science either have been refuted, or have been eliminated by economy and simplicity criteria applied to theories. It would be a mistake to suppose that they could have been refuted by facts alone, because even if we do not accept the strong theory-ladenness thesis of point 2, it would generally be agreed that facts are susceptible of a multiplicity of theoretical interpretations, and that if such value judgments were regarded as of overriding importance (overriding, that is, all except logic), the facts could have been accommodated, though perhaps at a cost to economy. But the requirement of theoretical economy or simplicity is not an adequate general answer either, for at least two reasons. First, what have been held to be *prima facie* simple theories have often been abandoned for more complex ones. Examples are: field theories in place of action at a distance, atomic theories in place of phenomenal volume and weight relations in chemical reactions, and Copernicus's heliocentric universe, which in his theory required more parameters than the geocentric universe it replaced. In most such cases, what was of overriding importance was not facts plus *prima facie* simplicity, but facts plus interpretation in terms of some intelligible or desirable world model. Secondly, no one has yet succeeded in presenting definitions of simplicity that are adequate for all the occasions on which appeal had been made to it. But it is at least clear that there is not one concept of simplicity but many, and the suspicion grows that simplicity is not in itself a final court of appeal but rather adapts itself to definition in terms of whatever other criteria of theory choice are taken to be overriding.

The most important of these other criteria in natural science is what I have called the *pragmatic criterion* of predictive success. In considering historical examples the question to be asked in philosophy of science is not so much what were the special local (social, biographical, psychological, etc.) factors at work in the immediate and short-term decisions of individuals, but rather whether there is any *general* criterion for the long-term acceptability of one theory rather than another, and for the replacement of old theories by new. This is to ask a question which presupposes that because all formal criteria such as verification, confirmation, and falsifiability seem to have broken down as criteria for theory choice in particular short-term scientific situations, therefore there are no general criteria of theory choice over the long term. But revolutionary accounts have not disposed of the objection that natural science, as well as being revolutionary in respect of *theories*, is also in some sense cumulative and progressive, and retains contact with the empirical world by means of long-term testing of theory complexes taken as wholes. If we press the question 'What is it that progresses?', the only possible long-term answer is the ability to use science to learn the environment, and to make predictions whose results we can rely on not to surprise us. It is this modification of the traditional empirical criteria of confirmation and falsifiability that I intend by the 'pragmatic criterion'.[3]

As successful prediction accumulates, the pragmatic criterion filters out both simplicity criteria and other value judgments. We can observe by hindsight that in the early stages of a science, value judgments (such as the centrality of man in the universe) provide some of the reasons for choice among competing underdetermined theories. As systematic theory and pragmatic success accumulate, however, such judgments may be overriden, and their proponents retire defeated from the scientific debate. Thus, the theological and metaphysical arguments against Copernicus, against Newton and against Darwin became progressively more irrelevant to science. This is not to say, of course, that *our own* preferences in choices between underdetermined theories are not themselves influenced by our value judgments and by beliefs which we take for granted, or that these will not be visible to the hindsight of future historians of science.[4] This is very likely to be so, but it does not conflict with the notion of accumulation of pragmatic success in science past, present, and future. There is also a sense in which value judgment enters into the very adoption of the pragmatic criterion itself —the judgment that the requirement of predictive success should override all other possible criteria of theory choice. This is the one value judgment that, of course, is *not* filtered out, but rather is presupposed in the pragmatic criterion. It is a judgment that has perhaps rarely been consciously adopted by any scientific society of the past, but it is one which, it is becoming increasingly apparent, may be consciously rejected in the future.

It is not my purpose here to discuss in detail the relation between what I have called the pragmatic criterion and more orthodox theories of objectivity and truth. But something more must be said to avoid misunderstanding. First, there is a difficulty about the notion of 'successful prediction'. If we were able to ignore the much-discussed difficulties referred to in point 2, and assume that there is a theory-neutral observation language for which there are clearly applicable truth criteria, we might be tempted to define 'increasingly successful prediction' in terms of an accumulating set of true observation statements deducible from the corpus of scientific theories. We cannot ignore these difficulties, however, or the consequent tendency to understand 'truth' not in a correspondence sense but as coherence within a given theory, and hence as theory-relative. Since theories of truth are themselves in considerable disarray, it is better to find some way of understanding 'successful prediction' independently of them. Here I suggest a pragmatic or ostensive appeal to the actual state of natural science since the seventeenth century, in which we can recognize an accumulation of successful prediction which overrides changing theories and is *independent* of particular conceptual schemes in which scientific successes are described in conflicting theories. The spaceship still goes, whether described in a basically Newtonian or relativistic framework. Pragmatic knowledge can be obtained without an absolutely theory-neutral descriptive language.

It may be illuminating to draw an analogy (only an analogy) between the method of natural science and the program of a computer designed to process environmental data and to learn to make successful predictions (for example, a character recognition device). The criteria of success of such a device can be made independent of the actual 'language' system used in the computer to store and process data and to give the orders for testing theories on more data. Equal success in two or more computers is consistent with their having widely different internal language systems, although of course some language systems may be more convenient than others for given kinds of data, and indeed there may be feedback mechanisms in the program which permit change of language to a more convenient one when this is indicated by the success and failure rate.

A second possible misunderstanding of the pragmatic criterion arises from the fact that technological applications are the most striking examples of accumulating successful prediction. Philosophers of science should perhaps disengage themselves from the Popper-induced prejudice that pragmatic application has nothing to do with the logic of science. On the other hand, successful prediction does not necessarily issue in technical control. Many theories enlarge our pragmatic knowledge (for example, about fossils, or quasars), without necessarily forming the basis of technology.

A third difficulty is that the relation between the pragmatic criterion and any theory of truth is obscure and needs much more examination than can be given to it here. But in the particular case of some kind of correspondence theory, to which philosophers of truth seem now to be increasingly drawn, there does not seem to be any *prima facie* conflict between such a theory and the pragmatic criterion. Current correspondence theories of truth tend to be expressed in terms of some relation of 'satisfaction' which holds between the world and true statements, and are in themselves independent of the question how such satisfaction is identified in particular instances. There are notorious difficulties about such identifications—the same difficulties that underlie the notions of underdetermined theories and criticisms of the basic observation language. The pragmatic criterion trades these difficulties for others by bypassing the question of the reference of theoretical language, and resting on the non-linguistic concept of successful prediction.

II

In considering whether natural science as defined by points 1 to 5 is an adequate model for the social sciences, we can add two further points:

6. There are not at present, and perhaps can never reasonably be expected to be, general theories in the social sciences that satisfy the pragmatic criterion of point 5—namely, theories that provide increasingly successful prediction and control in the social domain.
7. Moreover, since adoption of the pragmatic criterion itself implies a value judgment, it is possible to decide *against* it as an overriding goal for social science, and to adopt other value goals.

Point 7 does not presuppose the truth of point 6. I doubt if point 6 can be proved in any general way. On the actual present situation one can only observe what underlies complaints about the backwardness, theoretical triviality, and empirical rule-of-thumb character of most social science, in spite of limited success in establishing low-level laws in isolated areas. On the logical possibility, there have been attempts at general proof of the non-natural character of social science, attempts which derive from features of the social subject matter such as complexity, instability, indeterminacy, irreducible experimental interference with data, self-reference of social theorizing as part of its own subject matter, etc. I do not believe such proofs can ever be conclusive, if only for the reason that most of these features are also found somewhere in the natural sciences. If we use as an analogy for the method of natural science the computer which learns to predict its environment, an immediate consequence is that there will be some environments and some types of data which do not permit learning by any computer of limited capacity, for any or all of the reasons just listed. The social environment *may*, wholly or partly, be such an environment. I doubt if anything stronger can be said, and I doubt whether any attempt to formalize the situation further at this general abstract level is worthwhile. Satisfaction of the pragmatic criterion by particular social science theories needs to be argued case by case.

Point 7, however, remains. It is explicitly recognized in Marxist writings on the social sciences, and also in the older *Verstehen* tradition and in its more recent offshoot, hermeneutics (although the latter two traditions neglect the dimension of 'interest' that inevitably infects social theory according to Marxism). The rest of this paper will be devoted to exploring the consequences of point 7.

It is important to notice that point 7, together with points 1 to 5, imply a distinction between two sorts of 'value-ladenness' in social science. The first is analogous to theory-ladenness in natural science, and is the sense primarily in mind when empiricist philosophers have attempted to disentangle and exclude value judgments from a scientific social theory. It is the sort of value judgment that I have mentioned in point 4, which becomes associated with theoretical interpretations by virtue either of the selective interest of the investigator (for example in preferring to investigate stable systems as norms), or of adoption of those hypotheses which assert the world actually to be in some respects as it is desired to be (for example acquired characteristics either are or are not inherited according to preferred ideology). I have suggested that the crucial point about the natural sciences is that though such judgments may function heuristically in hypotheses, operation of the pragmatic criterion frequently filters them out, and how the world 'ought to be' frequently fails in face of how the world is, or rather in face of the only plausible and coherent ways that can be found of interpreting facts and successful predictions. Where the pragmatic criterion works in social science we shall expect some value judgments to be filtered out in a similar way—for example, it seems not impossible that currently controversial questions about the relationship, if any, between intelligence quotient and racial origin might be sufficiently defined to be made rigorously testable, and laws might be derived which satisfy the criterion of successful prediction. (Whether it would be *desirable* to adopt and try to rigorously apply the pragmatic criterion here is entirely another question.)

But where the pragmatic criterion cannot be made to work in a convergent manner it is not possible to filter out value judgments in this way. A second type of value judgment may then be involved, which in varying degrees *takes the place of the pragmatic criterion* in selecting theories for attention. These judgments will be *value goals* for science that are alternatives to the pragmatic goal of predictive success. Such alternative goals have often been recognized in the literature, for example by Weber in his category of value-relevance, and by Myrdal in arguing for explicit adoption of a value standpoint, preferably one that corresponds to an actual power group in society.[5] But alternative value goals have usually been recognized in the negative sense of the 'unmasking' of so-called non-objective biases, rather than in the positive sense of being consciously adopted goals other than the pragmatic criterion. It is difficult to make such standpoints conscious and explicit while they are operative, but the literature is now full of studies in the critical sociology of sociology, where the standpoints of the past, and of other contemporary groups of sociologists, are 'unmasked'. It is a well-known Marxist ploy to uncover the non-intellectual interests even of self-styled positivists: those who argue most strongly for a value-free and objective social science are shown to be those whose social and economic interest is in the status quo, and in not having the boat rocked by encouragement to explicit criticism and value controversy. And such studies are not found only in Marxist writers. Robin Horton, for example, has given an interesting analysis of the styles in social anthropology during this century in terms of the changing attitudes of the West, in its imperialist and liberal phases, towards its former colonies as they become politically independent and aspire to cultural autonomy.[6]

Weber carefully distinguished value-relevance from the value-freedom of the social scientist with respect to political action. That is to say, he accepted that judgments of interest select the subject matter of the human sciences, but denied that the social scientist as such should use his theories to argue any particular political practice. Even with respect to value-relevance, he argued that theories must ultimately be shown to be causally adequate. Thus Weber's own value-interest in studying, for example, the interrelations of capitalism and the Protestant ethic was doubtless to refute Marx's contention that the ideological superstructure is unilaterally determined by the economic substructure. But Weber insists that his theory of such relationships must be shown to be a factual theory of cause and effect, confirmable by positive instances and refutable by negative. Without going into the detail of Weber's discussions of methodology, it can I think be fairly concluded that he sees the goal of knowledge and truth assertion as essentially the same in the natural and social sciences, but that he has an oversimple view of the nature of causal laws in the natural sciences, which misleads him into extrapolating an almost naïve Millean method into the social sciences. He does not doubt that judgments of value-relevance are separable from positive science, and can in this sense be 'filtered out' of cognitive conclusions. Thus he has not yet made the 'epistemological break' involved in recognizing, questioning, and perhaps replacing the pragmatic criterion for social sciences, nor has he distinguished two sorts of judgments of value-relevance—those which can ultimately be eliminated by the pragmatic criterion and those which cannot because they depend on a view of causality that presupposes it.

There are others who have not understood the nature of this epistemological break. In a commentary on Myrdal's requirement of total explicitness of value standpoint and identification with some actual power-group, John Rex[7] finds an implied suggestion that objectivity inheres in the balance of power between such groups, and that this balance of power 'can be relatively objectively determined', as if what is is determined by the standpoint of the most powerful group. However, while it may be true that the most powerful group can to a greater or lesser extent impose its will upon the development of the social system, it does not at all follow that the theory informed by its value standpoint gives the true dynamical laws of that system on a pragmatic criterion, or the best theory on any other criterion except that truth resides in the barrel of a gun. Whether the unions or the sheikhs eventually gain control in Britain is irrelevant to the theoretical acceptability of either of their implied economic doctrines. And Christ and Socrates may have the best theories after all.

Myrdal himself is more careful, but he too leaves largely unexamined the exact relation between objectivity as sought in the natural sciences and the value criteria which are inevitably adopted in social science. Of science in general he writes:

> Our steadily increasing stock of observations and inferences is not merely subjected to continuous cross-checking and critical discussion but is deliberately scrutinized to discover and correct hidden preconceptions and biases. Full objectivity, however, is an ideal toward which we are constantly striving, but which we can never reach. The social scientist, too, is part of the culture in which he lives, and he never succeeds in freeing himself entirely from dependence on the dominant preconceptions and biases of his environment.[8]

If 'objectivity' in this sense is the ideal which is unattainable, then valuations are a necessary evil. Seen in such negative light, it is unlikely that the choice between valuations will be subjected to logical or philosophical scrutiny, and the vacuum is likely to be filled by power criteria or worse, in the manner of Rex. But if it is true that the ideal objectivity is unattainable and that valuations are necessary, the philosopher will surely be better advised to present this necessity in a positive light, and to critically examine the value choices that are then open. In points 4 to 7 I have attempted to articulate such a positive view by distinguishing value-laden theories, subject to the pragmatic criterion, from the value goals adopted for the total scientific enterprise.

Mannheim is another exponent of the sociology of sociology who has been misled by neglect of this distinction. As is well known, he adopts a relationism of total ideology according to which all knowledge (except logic, mathematics and natural science) is knowledge only in relation to some observer standpoint.[9] In our terms, this may be interpreted as a recognition of the value-ladenness of the social sciences, and the suggestion that some observer standpoint determines the criterion of evaluation. He then asks: Which standpoint is optimum for establishing truth? and goes on to reject the two classical Marxist answers—the proletariat, and the class-self-conscious Party subsection of the proletariat—and to put forward the intelligentsia, who, he claims, are powerless and interest-free, and understand the sociology of knowledge. Whatever be the merits of this particular choice, it is clear that Mannheim has now retreated from his fleeting glimpse of irreducible commitment to value goals and is asking for an ideal standpoint from which truth is seen in the same sense of truth as that appropriate to natural science. Apart from the falsity of the claim that the intelligentsia are disinterested, this leads Mannheim into the logical circle of sociology of knowledge that has often been discussed, namely: In relation to what observer standpoint is it *true* that the intelligentsia are disinterested? Whatever the answer, truth as thus defined is clearly still relational and not objective in the sense of the pragmatic or whatever other criterion is adopted for the natural sciences.

Of all writers on the sociology of knowledge, Alvin Gouldner perhaps comes closest to embracing point 7 explicitly. After a careful and devastating analysis of the sociological origins and determinants of American functional sociology, and a more brief analysis of Soviet sociology, he comes in his Epilogue ('The Theorist Pulls Himself Together, Partially') to the crucial question: What then are the sociological origins and determinants of Gouldner's unmasking exercise in respect of the American sociologists? He goes on explicitly to reject the approach of those methodologists 'who stress the interaction of theory and research . . . the role of rational and cognitive forces',[10] that is, the orthodox philosophers of science who reject value commitments. Rather, the sociologist must be *reflexive*—self-aware of his own place in his own standpoint—and must accept his involvement in it in a manner that requires a new *praxis*—a new lifestyle in which there is no ultimate division of himself as sociologist from himself as man. But more significant than this note of introspective moralizing is Gouldner's description of the sociologist's task in the following terms:

> Commonly, the social theorist is trying to reduce the tension between a social event or process that he takes to be real and some value which this has violated. Much of theory-work is initiated by a dissonance between an imputed reality and certain values, or by the indeterminate value of an imputed reality. Theory-making, then, is often an effort to cope with threat; it is an effort to cope with a threat to something in which the theorist himself is deeply and personally implicated and which he holds dear.[11]

Thus, 'the French Revolution, the rise of Socialism, the Great Depression of 1929, or a new world of advertising and salesmanship' are facts-as-personally-experienced, requiring not so much explanation in the sense of the natural sciences (which perhaps we can never have), as redescription (interpretation, understanding) in terms which make them cohere with a chosen order of values. One might compare with Weber's desire to rescue human ideals from dominance by substructures, whether economic or bureaucratic; Durkheim's sense of the need for social cohesion and stability in face of man's inordinate and irrational desires; the note of protest inseparably bound into Marx's 'scientific' concept of exploitation of man's labour power; and Gouldner's own quite unconcealed negative evaluation of the sociologies of Goffman and Garfinkel, whose origins he 'unmasks' and whose adequacy he judges not on grounds of a spurious 'objectivity' but on grounds of his own sense of the moral degradation of their pictures of the social world ('anything goes', 'espionage agents', 'demonic', 'camp', 'kicks', 'the cry of pain . . . is Garfinkel's triumphal moment', 'sadism').[12] In the light of Gouldner's rather clearsighted adoption of criteria other than the pragmatic, we may accept the challenge implicit in his statement early in the book that 'whether social theories *unavoidably* require and must rest *logically* on some background concepts [valuations] is a question that simply does not concern me here . . . this is a problem for philosophers of science'.[13] Though he appears here to remain agnostic, if his own analysis is acceptable it provides strong grounds for the methodological adoption of point 7, that is for the recognition that where the pragmatic criterion is inoperative, other value goals for social science should be self-consciously adopted.[14]

III

By way of conclusion let me rebut some possible empiricist misunderstandings, draw a consequence for the sociology of knowledge, and suggest an analogy for the choice of value goals.

It may be objected that, in emphasizing the need to make explicit choice of value goals for science as well as theoretical value assumptions, I have neglected the role that the pragmatic criterion (or indeed any realist criterion of truth that might be proposed) actually plays in the social sciences. To this objection I would reply that nothing I have said about the inapplicability of the pragmatic criterion or about the choice of value goals is intended to exclude the possibility that there are areas in the social sciences where the methods and criteria of the natural sciences are both workable and desirable. There are general laws of human behaviour (though I suspect only low-level laws rather closely circumscribed in domain), there are models and ideal types whose consequences can be explored deductively and tested, and there are limited predictions which are sometimes successful. Where these things are the case, we may speak of 'objectivity' in the social realm in whatever sense we wish to speak of it in the natural realm, and we *may* (not *must*) make the same choice of value goals for the social as for the natural sciences. What I am arguing is that it would be wilfully blind and neglectful of the responsibility of social science as a cognitive discipline to ignore the fact that much social science which is currently acceptable is not and probably never can be of this kind. I have been primarily concerned with the consequence that non-pragmatic value choices have to be made. There will of course be difficulties in demarcating one type of value choice from others, both where there are doubts about how far the pragmatic criterion can be taken and also about how far it should be taken, as in recent disputes, for example, about the racial inheritance of characteristics, or about whether Garfinkel should collect data that involve severe mental disorientation of people in their ordinary social relationships. No general rules can be given about such disputes, because they are themselves essentially value disputes about the goals of particular social researches. But all of this does not entail that there are no facts or laws in the social sciences, nor that where there are such, social theory should not be consistent with them. As in the natural sciences, social theories are *constrained* but not *determined* by facts. Whether we wish to extend the use of the concept of 'objectivity' beyond the domain of such facts to the recognition of value choice, as Myrdal does,[15] is partly a verbal matter.

A more fundamental empiricist objection, however, is the following.[16] It may be suggested, first, that there is great difficulty in actually articulating viable goals for social science which are alternatives to the pragmatic criterion, and second, that where such goals are apparently identified and described, their operation always in fact involves the pragmatic criterion. Traditionally various versions of the *Verstehen* thesis have been appealed to to provide alternative goals, but it seems that in any attempt to *understand* a person's behaviour, one is seeking to fulfil one's expectations about his future behaviour. Indeed, all human interaction depends on the success of some such predictions about mutual responsiveness, and this seems not unlike an application of the pragmatic criterion. This is surely correct, and it is not surprising to find successful fulfilment of expectations as a criterion in all reasoning about the world, including all lowest level inductive generalizations, whether about objects or persons. But this argument must not be made to prove too much. It certainly does not show that the pragmatic criterion as described in point 5 is sufficient to determine uniquely the theoretical interpretation of people's behaviour, for in this context we lack the wideranging and systematic generality characceristic of natural science, and consequently we lack what I have called the filtering-out mechanism that eventually eliminates value judgments as criteria of theory choice.

Secondly, a remark about consequences for the programme of the sociology of knowledge. Whether or not we wish to use epistemologically loaded terms like 'cognition', 'knowledge', 'objectivity', and 'truth' for acceptable theories in social science, a consequence of my arguments is that criteria of acceptability are *pluralist*—as pluralist as our choices of value goals. And if we wish to talk of *choice* of values it also follows that we presuppose a certain area of freedom in the activity of theorizing—we are not wholly constrained to adopt particular theories either by the facts, or by adoption of particular value goals, or by social and economic environment. Thus it would be inconsistent with the present thesis to hold a form of the sociology of knowledge according to which socio-economic substructure determines all forms ofknowledge, including presumably adoption or non-adoption of the sociology of knowledge itself. As has often been pointed out, such a determinist view, while not actually selfcontradictory, is somewhat self-defeating.[17]

But a weaker form of sociology of knowledge has been implicit in my presentation of the value basis of social science, because one of the grounds for holding that value choices are inseparable from social theories is precisely that other people's theories, and sometimes one's own, can be shown to be partially determined by social environment and interests. And yet if this 'can be shown' is taken in the sense of objective empirical knowledge, we are on the horns of the same dilemma: *either* here at least (and in the least likely place) we have got to accept a kind of sociology that is objectively empirical and interest-free, *or* it 'can be shown' only on the basis of yet another, and probably interest-influenced, value choice. The first horn of the dilemma is lethal, but the second is graspable just in virtue of the pluralist conception of value choice that has been generally adopted here. The question now becomes: Given that we are not *compelled* to adopt it, does the weaker sociology-of-knowledge thesis nevertheless commend itself to our value system as desirable—illuminative of dark areas in social interaction, and conducive to understanding of others and of ourselves? The question can only be answered by consideration of particular examples, and in terms of one's personal reaction to them. I would answer for myself that many examples in the work of Marx, Mannheim, Myrdal, Gouldner, and some current 'critical sociology'[18] do seem to be thus illuminating.

Finally, a convenient analogy. I suggest that the proposal of a social theory is more like the arguing of a political case than like a natural-science explanation. It should seek for and respect the facts when these are to be had, but it cannot await a possibly unattainable total explanation. It must appeal explicitly to value judgments and may properly use persuasive rhetoric. No doubt it should differ from most political argument in seeking and accounting for facts more conscientiously, and in constraining its rhetoric this side of gross special pleading and rabble-rousing propaganda. Here the inheritance of virtues from the natural sciences comes to the social scientist's aid, and I hope nothing I have said will be taken to undermine these virtues. The fact that the view of the social sciences presented here is more often associated with the particular choice of value goals of the revolutionary left[19] does not in the least invalidate the general argument, nor reduce—rather, it increases—the need for the moderate centre and right to look to its own value choices. Neither liberal denial that there are such value choices nor cynical right-wing suppression of them from consciousness will meet the case.

Hume attempted to divorce the question of truth from that of value, while certain scientific humanists have attempted to derive value from truth. A consequence of my argument on the other hand has been that, at least in the sciences of man, a sense of 'truth' that is not merely pragmatic may be derivable from prior commitment to values and goals.

Notes

1 I should like to express my thanks to those who commented on a first draft of this paper at the meeting of the Thyssen Philosophy Group at Ross-on-Wye, 26–28 September 1975, and also to David Thomas for discussions on the general problem of values in social science.

2 I have discussed these matters in *The Structure of Scientific Inference*, London, 1974, chs 1 (reprinted as ch. 3 above), 2 and 12, and in 'Truth and the growth of scientific knowledge' in *PSA*, 1976, ed. F. Suppe and P. D. Asquith (Philosophy of Science Association, East Lancing, Mich., 1976) (reprinted as ch. 6 above). Since the notion of 'underdetermination' has been exploited particularly by Quine, I should like to say that I do not accept his distinction between 'normal scientific induction' and 'ontological determinism', according to which it seems to be implied that purely scientific theories can eventually be determined uniquely by inductive methods. Some of my reasons for this rejection will emerge in this chapter.

3 It was Duhem's holist account of theory-testing in *The Aim and Structure of Physical Theory* first published as *La Théorie physique*, 1906, English edn, Princeton, 1954, which foreshadowed they demise of later and narrower criteria of empirical test. The work of I. Lakatos has more recently familiarized philosophers of science with the problem of theoretical acceptability in long-term historial perspective, although his criteria for 'progressive research programmes' do not include the predictive aspects of the pragmatic criterion adopted here. See particularly his 'Falisfication and the methodology of scientific research programmes' in *Criticism and the Growth of Knowledge*, ed. I. Lakatos and A. Musgrave, Cambridge, 1970.

4 Recent studies in the history and sociology of natural science indicate that there has been far more influence upon theories from evaluations and non-scientific standpoints than has generally been realized. See for example P. Forman, 'Weimar culture, causality and quantum theory, 1918–1927: adaptation by German physicists and mathematics to a hostile intellectual environment', in *Historical Studies in the Physical Sciences,* vol. iii, ed. R. McCormmach, Philadelphia, 1971; papers in *Changing Perspectives in the History of Science*, M. Teich and R. M. Young, London, 1973; and many references in Barry Barnes, *Scientific Knowledge and Sociological Theory*, London, 1974, and David Bloor, *Knowledge and Social Imagery*, London, 1976, and in ch. 2 above.

5 M. Weber, *The Methodology of the Social Sciences*, ed. E. A. Shils and H. A. Finch, New York, 1949, and *The Theory of Economic and Social Organization*, Oxford, 1947, ch. 1. The second part of *Methodology* and ch. 1 of *Theory* are reprinted in M. Brodbeck, ed., *Readings in the Philosophy of the Social Sciences*, New York, 1968. G. Myrdal, *The Political Element in the Development of Economic Theory*, London, 1953, *Value in Social Theory*, London, 1958, and *Objectivity in Social Research*, London, 1970.

6 R. Horton, 'Lévy-Bruhl, Durkheim and the scientific revolution' in R. Horton and R. Finnegan, eds, *Modes of Thought*, London, 1973. For another unmasking of positivism see A. Gouldner, *The Coming Crisis of Western Sociology*, London, 1970, esp. ch. 4.

7 J. Rex, *Key Problems of Sociological Theory*, London, 1976, pp. 164–6.

8 Myrdal, *Value in Social Theory*, p. 119.

9 K. Mannheim, *Ideology and Utopia*, London, 1936, esp. chs 2 and 5. It

is of course ironic that it is the introduction of ideological and social criteria into the interpretation of *natural* scientific theories, by Kuhn, Feyerabend and others, that has played a large part in the revival of contemporary debate about sociology of knowledge.

10 Gouldner, *op. cit.*, p. 483.

11 *Ibid.*, p. 484.

12 *Ibid.*, pp. 378–95.

13 *Ibid.*, pp. 31–2.

14 Another writer who makes explicit the choice of value goals for knowledge is J. Habermas, especially in Appendix to his *Knowledge and Human Interests*, London, 1972. Habermas there makes a threefold distinction: the *technical* interest of the empirical sciences; the *practical* interest of the historical sciences, defined as 'the intersubjectivity of possible action-orienting mutual understanding' (p. 310), somewhat in the *Verstehen* tradition; and the *emancipatory* interest of the social sciences, whose function is critique of the established social order. In the light of examples such as are quoted in n. 17 below, this is to put matters into a rather too restricted straitjacket.

15 Primarily in his later *Objectivity*, pp. 55–6, where he supplements the positivist view of objectivity implied in *Value*, p. 119 (quoted above), in the following terms: 'The only way in which we can strive for "objectivity" in theoretical analysis is to expose the valuations to full light, make them conscious, specific, and explicit, and permit them to determine the theoretical research.'

16 This objection was raised by participants in the Ross-on-Wye Conference, 26–28 September 1975.

17 For a discussion of this see my 'Models of method in natural and social sciences', *Methodology and Science*, vol. viii, 1975, 163, and ch. 2 above.

18 There are some good (and some bad) examples in R. Blackburn, ed., *Ideology in Social Science*, London, 1972. C. B. Macpherson describes the function of social theory as *justification* of a social system (*ibid.*, pp. 19, 23—cf. Gouldner's 'reduction of tension' quoted above); M. Shaw, on the other hand, in a critique of Gouldner, takes the goal to be the overcoming of academic sociology 'in the development of the revolutionary self-consciousness of the working class' (*ibid.*, p. 44). In more moderate vein, in I. L. Horowitz, ed., *The New Sociology: essays in social science and social theory in honour of C. Wright Mills*, Oxford, 1964, S. W. Rousseas and J. Farganis assert that ideology (inseparable from social theory) 'must be concerned with the human condition and its betterment in an always imperfect world. Its justification for being is, in a word, progress' (p. 274).

19 It is significant that G. S. Jones, in an Althusserian piece in Blackburn's *Ideology*, denies that it is values or interpretations that are involved in theory choice, and holds that what is required is new concepts of structure (p. 114). In other words, the hard Marxist reverts to a view of the social sciences as theory-laden only, but of course the theory (and the values) are those of Marx.

9 *Habermas' Consensus Theory of Truth*[1]

Critical science and the theory of interests

The question of truth is central to current discussions in both of the major contemporary styles of philosophizing. In the Anglo-American linguistic and empiricist tradition there is a lively response (some might say backlash) to apparent difficulties caused by recent recognition of theory change and meaning variance in science. And within the continental hermeneutic tradition there is raised the central question of the truth status of interpretations in the cultural sciences where these appear not to be subject to the criteria of empirical science. Let me say straight away that I believe that the almost universal dependence on versions of the correspondence theory of truth among analytical philosophers will prove seriously inadequate to both forms of the epistemological problem, and that we have to face here a deep challenge to many entrenched assumptions of empiricism—assumptions that are too infrequently brought to the light of day in these discussions.[2] We still need to consider two questions which may look old-fashioned, but which have never been straightforwardly answered, and which gain new relevance from contemporary developments. They are

1. Is knowledge, as empiricists and positivists would have it, restricted to the empirical and the logical, and correspondingly, is 'truth' only appropriately predicted when the linguistic circumstances are empirical and logical?
2. If not, then can we attain new concepts of knowledge and truth that do justice to interpretations both in theoretical natural science, in social science and in ethics, while not departing too far from traditional understandings of these concepts?

These questions are posed in the idiom of liberal empiricism. But it is significant that the Marxist tradition in regard to natural and social science shows two opposing tendencies which are not unlike the 'liberal' distinction between empiricist and hermeneutic methodology. Marx's own writings are ambiguous, but, as interpreted by Engels and continued in the 'hard' Marxism of the Soviet Union and associated Communist Parties, and in the Althusserian school of French philosophers, Marxism emphasizes the empiricist tradition and hopes for its extension into the social sciences. This is the Marxism that claims that it itself is a 'scientific' theory of the world, and that the laws of conflict and development it finds in history have a status similar to that of the causal laws of natural science. This view has provided a justification for the subordination of certain scientific debates in the Soviet Union to Marxist theory in the form of dialectical materialism, notably the debate in genetics surrounding Lysenko.

Since the publication of the works of the 'young Marx', however, and associated with the 'human' Marxism of Lukács and the subsequent New Left outside the Soviet Union, there is an opposing tendency to regard the empiricist tradition in science as necessarily linked with the capitalist-industrial societies of the Western World, and therefore with support of the status quo and of inhumane exploitation of the peoples of both the Western and the Third Worlds. Detached value-free science becomes subject in this view to the same criticisms as those already brought against the empirical philosophies of science by such Western writers as Feyerabend.

One of the main sources of this Marxist-inspired critique has been the Frankfurt school of social philosophers to which Marcuse and Habermas originally belonged. Marcuse has suggested that the establishment of socialist society will change even natural science, since empiricism is after all *not* neutral in the ideological struggle, but has been allied with bourgeois capitalism. Habermas, however, has rejected this view on the grounds of his own understanding of natural science as predictive and technical—the *instrumental* aspects of natural science do not change in the liberated society, what changes is people's attitudes towards science and its application. Only in this sense is empirical science value and ideology laden.[3]

In the course of development of his philosophy Habermas provides one of the most perceptive and far-reaching analyses of truth in contemporary discussion. I shall devote this chapter to exposition and consideration of his theory of truth as so far available in his published writings. He has often been regarded by analytic philosophers as an obscure and inaccessible writer, and his philosophy is indeed a wide-ranging and not always closely argued structure, originating from many different problematics in German and English-speaking traditions. Although his concern with a theory of knowledge arises from the largely German context of *Natur-* versus *Geisteswissenschaften*, his recent more mature ideas seem to have developed principally in the context of Anglo-American linguistic philosophy and philosophy of science. It may help to commend his theory to the English-speaking world if he is interpreted as having arrived by a somewhat different route at many of the problem situations now dicussed in the empiricist tradition, and indeed he has recently been stimulated to explicate his theory in the context of some familiar problems of analytic philosophy of science.

Habermas' primary thesis is that the positivist theory of science is inadequate because it does not take account of what he calls *communicative* knowledge, that is, of the linguistic conditions of interpersonal communication, and because it is incapable of self-reflection, that is of applying itself to itself. His answer to our first question about the positivist restriction of knowledge to the empirical is therefore negative, and he arrives at this answer by critical adoption of some of the recent theses of philosophers of science themselves, especially those of Popper and Kuhn. He uses these to build up his own account of the 'empirical-analytic' sciences, which are not exactly co-extensive with the natural sciences, because the social sciences will also be found to have empirical-analytic components.[4]

In *Knowledge and Human Interests* (hereafter *KHI*) the model for the empirical sciences is a modification of that of Charles Peirce. Science is a necessarily cumulative process, defined by the goal of successful prediction which is ensured by its method of feedback control. This is to speak in terms of what Habermas later distinguishes as *communicative action*—it is the naïve and unreflective action of the scientific community in following the tacit rules of methodology of science. When we reflect on the justification of the *statements* scientists thereby produce, however, we rise to the level of logic of science or epistemology, or what Habermas calls *discourse*. At this level, questions arise about truth and reality, and here Habermas finds that Peirce's account is ambiguous. On the one hand Peirce defines 'truth' as the ideal permanent consensus of scientists at the limit of the application of their method of testing and self-correction, and defines 'reality' as the totality of possible true statements. On the other hand Peirce wants to retain some notion that this set of true statements 'corresponds' to an external reality which somehow guarantees the success of science by self-correction or 'survival of the fittest'.

Habermas points out that Nietzsche had already shown that it is 'possible to conceive a reality that can be resolved into a plurality of fictions relative to multiple standpoints'[5] —that is, in more recent terminology, that theories are underdetermined by reality. Hence, Habermas concludes, Peirce's contemplative notion of reality uniquely constraining the truth-seeking process of science dissolves, and he is left only with the instrumental success of science in the past, and the consensus which is the ideal of the scientific community in seeking technically adequate theories in the future.

Habermas therefore adopts Peirce's second alternative, of developing an explicit consensus theory of truth. In *KHI*, however, he does no more than hint at this, preferring rather to contrast the mode of knowledge which constitutes the empirical sciences with that of the hermeneutic sciences. This leads him, in the Appendix of *KHI*, to his well known threefold analysis of knowledge determined by three types of universal human *interest*. First, the 'technical interest' in mastering nature constitutes knowledge as successful prediction and dictates the methodology of the empirical sciences. Second, the 'practical interest' in free communication between persons constitutes knowledge as interpretation of meanings, that is, as hermeneutic, and dictates the methodology of history and the cultural sciences insofar as they aim at understanding. (English speakers should note that 'technical' and 'practical' as here used by Habermas are contrasted in the Aristotelian sense of *techne* and *praktikos* where *praktitos* is concerned with human affairs or politics, not as contrasted with theory in application both to nature and society.)

The third mode of knowledge is constituted by the 'emancipatory interest' in liberation from constraints of all kinds, both of natural necessity and of social domination. It becomes clearer in Habermas' later writings that this third mode is derivative from the other two, and depends on deviance from the ideal conditions of knowledge in the other two modes, since the emancipatory interest issues in the critique of social and psychological domination in general and of distortive ideologies in particular, which in the conditions of our society prevent the free pursuit of the technical and practical interests.[6] In other words, although there is a tidy correspondence between the three interests and three current types of sociological method, that is empirical, hermeneutic and critical, the threefold symmetry is misleading. There are basically only two modes of knowledge, and the empirical and hermeneutic both have to become self-reflective and critical to emancipate themselves from constraints that do not belong to their proper goals.

The upshot of *KHI* for a theory of truth was the appearance of an exclusively instrumental interpretation of empirical science, together with quite unrelated and still rather mysterious claims for the 'objectivity' of hermeneutics 'constituted by' the domain of communicating persons in place of that of natural objects. This appearance of instrumentalism was partly due to an absence in *KHI* of analysis of the significance of scientific *theory*, except for some dismissive remarks early in the Appendix which seemed to imply that belief in the truth of theories is an illusory legacy of Greek *theoria* and of a contemplative and catholic view of man's relation to nature. Two critical questions therefore arise from this stage of Habermas' discussion:

1. How are we to understand the *theoretical* dimension of empirical science, which appears to be redundant to its instrumental success, and yet does still seem to present itself to scientists as an indispensable 'search for truth'?
2. How can concepts of 'truth, 'objectivity', 'knowledge', be relevant at all to hermeneutic science, when there appears to be no objective constraint here analogous to the

correspondence between prediction and the world as ensured by the test method of empirical science?

Habermas' answer is essentially to dissociate 'truth' from 'correspondence' altogether, and to locate it in consensus both in the empirical and the hermeneutic sciences. This theory is developed in what he calls a 'theory of communicative competence'. I shall give an account of this under three headings: the critique of scientism, the ideal speech situation, and the truth status of hermeneutic discourse.

The critique of scientism

By 'scientism' Habermas means two things.[7] First, the view that empirical science is co-extensive with knowledge, and is adequate for knowledge of persons and societies as well as things. Second, scientism implies the view that empirical knowledge is sufficient for its own explanation. His arguments against the first thesis follow lines made familiar by Winch and others, namely that knowledge of persons and societies involves interpretations of meaning implicit in human language and social institutions. Moreover Habermas adds the thesis that the *interest* of hermeneutic science lies in such interpretive understanding with the aim of interpersonal communication, and not (or not exclusively) in the empirical interest of prediction and successful test. These arguments are still presupposed in his more recent work, although the notion of 'interest' is now subordinate to a notion of 'communicative competence'.[8] This is a result of the distinction he now wishes to make between the levels of unreflective *action*, where interests both technical and practical are pursued according to socially induced norms and practices, and reflective *discourse*, where we are concerned with propositions, and where justifications of the validity of actions takes place. Communicative competence, unlike action and interests, is therefore *linguistic*, and refers to our ability to argue the validity of what is unreflectively done by scientists, historians, etc. in pursuit of their respective knowledge-constitutive interests. This distinction need not imply of course that the two activities of action and discourse are always kept separate—they may be conducted by the same person at the same time—at any point anyone may be his own epistemologist. But it is important for Habermas that they be conceptually separated.

Against the thesis that empirical knowledge is self-explanatory, Habermas argues that no empirical evolution-type theory of science or society can explain the emergence of the relevant norms presupposed in social institutions, because such theories themselves will rest on the prior introduction of norms and hermeneutic viewpoints. This is a special case of an argument for the intrinsic value-ladenness of social science, that is, the thesis that comprehensive social theories cannot avoid adoption of particular value standpoints that are not decidable by empirical data. The thesis of the value relevance of the social sciences is familiar from the most usual interpretation of Weber's work, which seems here to be accepted by Habermas. As an example, Habermas refers to Chomsky's easy extrapolation of the data into a theory of the universal innateness of certain structural linguistic competences, when these have actually been shown to be universal only with respect to some localizable cultures.[9] Habermas traces what he regards as this illicit extrapolation to the fact that Chomsky makes into a global and universal scheme our local scientific naturalism and the rational organization of our society. Once this local consensus is questioned, however, the notion of a correspondence between Chomsky's linguistic model and the world is seen to be an artefact of late capitalist naturalism and scientism within which the grounds of correspondence truth do not come into question. Correspondence truth is plausible only within a single conceptual framework, and this has been undermined both by recognition of radical theory change in empirical science, and by recognition of the value relevance of social theories.

Nevertheless the notion of correspondence is important for scientific knowledge; the artefact stands for something, and if not for the concept of truth, then it must be redescribed otherwise. The way Habermas redescribes it indicates at the same time his reply to the objection that his theory of empirical science is wholly instrumental, and implicitly gives his answer to the problems of theory ladenness and meaning variance. Let me quote an important but rather opaque paragraph:

> The *objectivity* of experience could only be a sufficient condition of *truth*—and this is true of even the most elementary empirical statements—if we did *not* have to understand theoretical progress as a critical development of theory languages which interpret the prescientific object domain more and more 'adequately'. The 'adequacy' of a theory language is a function of the truth of those theorems (theoretical statements) that can be formulated in that language. If we did not redeem these truth claims through argumentative reasoning, relying instead on verification through experience alone, then theoretical progress would have to be conceived as the production of new experience, and could not be conceived as reinterpretation of the *same* experience. It is therefore more plausible to assume that the objectivity of experience guarantees not the *truth* of a corresponding statement, but the *identity* of experience in the various statements interpreting that experience.[10]

What Habermas is saying here I think can be paraphrased in five points.

1. Since even the most elementary observation statements are expressed in terms of some theory language or other, and since these theory languages change with time, truth cannot inhere in observation statements simply as correspondence between statement and the empirical world.
2. We therefore have to understand theory languages not as directly describing the world, but as *interpreting* it more and more 'adequately' as science develops.
3. 'Adequacy' is measured by experimental verification, but also necessarily by argumentative reasoning from the truth of theoretical postulates formulated in the language.
4. If adequacy were measured by verification alone we should fall into the meaning variance problem, because there would be no linguistic means of identifying the experiences expressed in the language of one theory with those expressed in the language of another. (It has to be assumed here either that ostensive face to face identifications do not work, or that they cannot be assumed to be available because communication between scientists typically takes place linguistically at a distance from the actual experiments referred to—as Habermas puts it in a different context, science is 'dialogical', not 'monological'.)

5. Therefore, in order to guarantee the identity of reference of observation statements made in different theoretical languages which are 'about' the same subject matter, we cannot rely on their 'correspondence' with the subject matter, but we rather need communication and argumentation between and within different theory languages.

This explains the at first puzzling insistence of Habermas that, far from advocating an instrumentalist theory of science, he regards this as a bad misunderstanding, for the argument shows that it is recent insights about theory change through the history of science that finally makes instrumentalism untenable as an account of discursive scientific *knowledge*.

Habermas' resolution of the problems of instrumentalism and theory change, as paraphrased in these five points, leaves some questions unanswered. He does not explicitly tell us *how* conversation across theories is possible, even though for him this is the primary justification for having theories at all. And a corollary of this is that he needs to explain the notion of the truth of theoretical postulates in a way that is not tied to correspondence or to merely local consensus. With regard to the question of cross-theory communication, there are hints that he would give a historical interpretation of the significance of theory languages. With Gadamer he recognizes that a language community is necessarily a bearer of *tradition*. In reply to those pluralists (Wittgenstein, Winch, Feyerabend) who see the language games as wholly independent systems, Habermas maintains that the problem of meaning variance is a pseudo-problem. Every natural language contains the resources for understanding other natural languages, including its own past. Applied to the sequence of scientific conceptual frameworks this means that the theoretical languages cannot be constructed totally de novo and independently of theoretical tradition. But it must not be forgotten that tradition, and therefore theory, is a matter of interpersonal communication, not of correspondence truth.

Let us return to the question of truth, however, and consider Habermas' alternative to the correspondence theory in terms of an ideal consensus.

The ideal speech situation

The truth of utterances in both empirical science and in hermeneutic interpretations is to be understood as the ideal consensus of competent practitioners of those disciplines. This leads immediately to the problem that afflicts all consensus theories of truth—how is the 'ideality' and the 'competence' to be described so that false consensus can be distinguished from true? Habermas argues that the ideality is to be found in the transcendental conditions of discourse as such, and the competence is to be found in the transcendentality of the mode of knowledge, that is in those conditions that make it the interest based knowledge it is.

Let us take the ideal speech situation first. As they actually appear in history, discourse and argumentation are intimately involved in what Habermas calls, in Marxist terminology, domination or authority structures (*Herrschaft*). That is, they are distorted by interests, by external oppression and repression, and by the psychic hang-ups of participant individuals. It is the function of both critique of ideology and psycho-analytic therapy to unmask these and to liberate discourse from them. Unlike theorists of 'total ideology' such as Karl Mannheim, Habermas does not believe that the notion of interest-free discourse is itself illusory or necessarily interest-based. In engaging in discourse, participants are aiming to detach themselves, at least momentarily, from action and interest, and are committing themselves to the assumptions that they are accountable for the validity of their utterances, and that the function of argument and mutual critique is to arrive at truth. At least a necessary condition for this is that in an ideal discourse consensus would be reached. One may compare with Francis Bacon's programme for stripping away the idols of the mind so that scientific induction can operate in a *tabula abrasa* free from prejudice induced by irrelevant social and individual distortions. Like Bacon, Habermas gives concrete content to the character of the idols, or ideologies, but unlike Bacon, he is not so specific about the method then to be pursued in valid argumentation.

Thomas McCarthy describes the informal conditions for Habermas' ideal speech situation as follows: '[Habermas'] thesis is that the structure is free from constraint only when for all participants there is a symmetrical distribution of chances to select and employ speech acts, when there is an effective equality of chances for the assumption of dialogue roles.'[11]

Particular symmetrical chances follow from this for various classes of speech acts: all participants must have equal chances of engaging in discourse and of putting forward justifications, refutations, explanations and interpretations. They must have equal chances of sincerely putting forward their own inner feelings and attitudes, and they must have equal status with regard to the power to issue permissions, commands, etc. All this presupposes that the ideal speech situation can take place only in conditions of an ideal form of life. The attainment of truth is not independent of attainment of conditions of freedom and justice.

The other important characteristic of knowledge as reached by consensus is its *universality*. Like Kant and Popper, Habermas adopts this as a regulative principle for the propositions that express empirical knowledge. Independently, and more crucial for his argument, he asserts the implicit universalibility of empirical and practical discourse over all possible participants, as would be ensured by the symmetry conditions. In *Legitimation Crisis* this principle is used to solve the relativist problem of plurality of viewpoints: 'argumentation is expected to test the *generalizability* of interests, instead of being resigned to an impenetrable pluralism of apparently ultimate value orientations. . . . Only on this principle of universalization do cognitivist and noncognitive approaches in ethics part ways.'[12] Thus in ethics as in every other cognitive activity, consensus based on the interests and agreement of only a limited speech community is generally false consensus. True consensus by definition requires the indefinite extendability of the ideal speech community, presumably throughout space and history. And it is Habermas' claim that this extendability in itself is sufficient for a discourse to contain truth claims.

The conception of the ideal speech situation is certainly very strongly counterfactual. What then is its status? It is not something that is empirically realized in history, and perhaps is never realizable. On the other hand it is not an arbitrary game. Habermas claims that it is a transcendental condition of commitment to discourse, and like the kingdom of heaven for Christian believers, it is *anticipated* by sincere participants in discourse. Such commitment and anticipation *are* found in history, and moreover in Habermas' view they constitute one of the highest functions of evolution. The ideal may be counterfactual, but 'on this unavoidable fiction rests the humanity of intercourse among men who are still men'.[13]

Habermas often refers to Popper's method of falsifiability and corroboration for the empirical sciences, and some of his discussion of consensus in the hermeneutic sciences seems to echo this language, so it may be illuminating to compare the 'anticipation' of the ideal speech situation with Popper's account of the ideal 'true' scientific theory. That an ideal speech situation occurs is *falsifiable* but not verifiable, like the occurrence of a true theory for Popper. Like Popper's true theory, it may never be realized in history, and indeed it seems as though we could not certainly recognize it if it were. Like Popper's true theory, it is a regulative ideal, presupposed in the decision to enter a certain form of life, that is, the scientific community of rational discourse. The parallel is indeed closer than this, because an account of empirical science something like Popper's is incorporated in Habermas' notion of the ideal speech situation, and the justification of natural science for him is itself an application of the regulative ideal. But Popper (like Peirce in his natural selection theory of truth) also maintains that truth is to be understood as correspondence with reality, so that there is some kind of asymptotic relation between the world and the sequence of theories proposed, corroborated and falsified. Habermas, on the other hand, rejects the correspondence account, and locates truth in the consensus which the scientific community would reach at a postulated limiting point of this process.

Some of the difficulties inherent in Popper's and Peirce's use of the terminology of *limits* here (including Popper's misleading 'verisimilitude' or 'approach to truth'[14]) infect also Habermas' conception. For it has often been pointed out that in examining the sequence of theories even in empirical science we cannot find any rules that would justify us in supposing that the propositional expressions of theories are asymptotically approaching the truth. This is even true of such simple hypotheses as those which postulate the proportion of a given property in a mixed large population on the basis of samples from that population (which is Peirce's paradigm case). *A fortiori*, theories about complex natural domains which claim generality on the basis of finite evidence can never be guaranteed to be 'approaching the truth' in any sense that permits us to find an accumulation of true or approximately true general statements. It was precisely recognition of this fact that was forced upon us by the accounts of theory change as revolutionary and not cumulative. There is no accumulation of true general theoretical propositions as such, only of pragmatic empirical learning which is not propositional, and of the stock of true statements in terms of which *we*, from the standpoint of our accepted theories, interpret the science of the past. The limiting process can never be an anticipation to be actualized in the future, only an interpretation based on past success.

This is a point which Habermas recognizes in relation to Peirce. What then is his notion of the anticipated consensus with regard to empirical science if it is not a version of Popperian realism? Although the striving for consensus *now* is necessarily dependent on the communicated tradition of the past sequence of theories, I do not think that the anticipated ideal consensus can be understood in terms of an actual sequence of theories in the future. It is rather an ideal that is equally relevant to all points of time and, what is more important, to all temporary conceptual frameworks. For in terms of the dialectic between Popper and Kuhn, Habermas accepts the Kuhnian insight that actual languages and conceptual schemes change, but against Kuhn he does not accept that this results in a relativism of truth; for within the domain of empirical science, truth is a demand and a commitment now, which must entail abstraction from local *interests*, but cannot involve abstraction from the particularity of the local conceptual schemes in terms of which true propositions must be expressed. We may usefully compare with Habermas' discussion of 'ideal history' in his review of Gadamer's *Truth and Method*.[15] Here he quotes Danto's view that historiography is a necessarily incomplete enterprise, because every historical narration incorporates judgments derived from *subsequent* events, including the 'meaning' of the narrated events in the context for which the historian himself writes. There cannot be any ideal 'last historian', according to Danto, because the historian's own writing of history is itself a historical event, the progress of which will itself in principle have to be written into the significance of all narrated past events, and so on. On the contrary, says Habermas, *every historian is his own last historian*: every historian *anticipates* the future in order to complete his story; every historian includes a viewpoint on the nature and destiny of the world.

The point becomes more intelligible in the light of the scientific parallel. Every theory making truth claims in a particular conceptual framework includes its own 'anticipations' of the total nature of the world as far as it is relevant to that theory. The commitment to anticipated consensus is the commitment to abandon falsified positions, and also to abandon conceptual schemes that do not lead to consensus. There is no last theory or theorist in the sense that science stops there, forever frozen in whatever conceptual scheme happens to be then current. But every serious theory and sincere theorist is 'the last', in the sense that *that* is where the accountability in the face of ideal consensus operates for him. To enter the scientific community presupposes acceptance of that accountability.

The parallel between scientific theory and hermeneutic viewpoint can be made close, but there still remains a tension in Habermas' thought in both cases between truth defined in terms of consensus in the ideal speech community on the one hand, and the recognition of changing conceptual frameworks on the other. It is a tension Habermas goes some way to resolve in his insistence that truth is not a predicate of propositions, but a predicate of claims made in speech acts. It seems that we have to move very far from correspondence truth towards a conception in which truth is multiply-related: it is a function of the total circumstances of the ideal speech situation, which will vary with the culture of a given society and therefore with its conceptual frameworks, although it will not vary with respect to the symmetries and universalizability demanded of ideal speech.

Many other questions about the ideal speech situation remain, of which I can mention only a few. First there has been much debate, into which I shall not enter, about the possibility of abstracting discourse from interests. Habermas' argument here is perhaps too reminiscent of Mannheim's for comfort, at least for the comfort of Marxists. Where Mannheim identified the *intellectuals* as the privileged class which was interest and therefore ideology-free (that is his *own* class!), Habermas identifies the *intellect* itself as the privileged human function. All men are supposed to share the distinctively human capacity for discursive self-reflection that constitutes ideal speech. Nevertheless, one may wonder both about the political and managerial consequences for ideal universal participation in decision-making that Habermas draws from this, and also about the high human significance given to argumentation and rational justification as compared with other possible contenders, for example intuitive, artistic and spiritual faculties. Morever, Habermas seems to have an unduly optimistic and Enlightenment-oriented view of human rationality. This is connected with his almost naturalistic tendency to evaluate this unique product of evolution as thereby necessarily its highest product and necessarily good.

To be fair to Habermas, the description 'intellectual' does not quite do justice to his conception of ideal speech. For this is explicitly said not to be confined to the rules of formal logic, and it depends on competence in communication about norms (for example, in moral discourse) as well as competence in descriptive discourse. But this raises further questions. What are the criteria of competence in discourse, whether technical or practical, if they are not or not only found in the canons of formal logic? Here Habermas' remarks are admittedly programmatic. He refers to the linguistic theories of Strawson, Austin and Searle, and the discussions of non-formal argument in Toulmin, to introduce what he calls a 'universal pragmatics' which will take account of the forms of rational but non-deductive argumentation such as occur in actual occasions of discourse.

More important than the attempted details of a universal pragmatics, however, is the question whether the whole conception of commitment to ideal discourse can be abstracted from cultural points of view. We have seen in the cases of historical narrative and scientific conceptual framework how the ideal is an *end* of discourse not in the sense of the final goal of time-based sequences, but of the doubtless counterfactual *telos* of discourse at every point of time. Even so, the implication is that *every* culture implicitly contains this very ideal of truth. Habermas wishes to describe this as neither an empirical description, since it is highly counterfactual, nor an option for arbitrary decision, but a transcendental implication of discourse as such. But what is his response to actual situations in which it is of the essence of the culture *not* to recognize such an ideal? These may be cases of totalitarian oppression in complex societies, or mythopoeic authority in archaic ones. His response is the moving but nevertheless ungrounded *judgment* 'on this unavoidable fiction rests the humanity of intercourse among men who are still men'. Whatever may be said about the transcendental justifiability of more specific evaluations, this presupposition about discursive argument seems for Habermas to be an ultimate value judgment.

The truth status of norms and evaluations

This brings us to the crucial question of the truth status of norms and evaluative judgments in general. Habermas uses the term 'decisionism' to refer to the view of norms and goals as ultimately ungrounded which is common to social theorists such as Weber, Parsons and Luhmann. For Weber and Parsons there is an irreducible dualism between, on the one hand, the ultimate values and goals of action (which in Parsons' terminology are 'arbitrary'), and on the other hand the facts, together with rational decision procedures which can be justified by optimizations of the facts. For Luhmann, as paraphrased by Habermas:

> It is meaningless to probe behind the factual belief in legitimacy and the validity claims of norms for criticizable grounds of validity. The fiction that one could do so if necessary belongs to the constituents of reliable counterfactual expectations. These, in turn, can be comprehended only

from a functionalist point of view, that is, by treating validity claims as functionally necessary deceptions.[16]

For Habermas, however, the claim within discourse that it is possible to argue about, justify and reject claims for the validity of norms is not a socially expedient deception, but is the very meaning of argumentative discourse as such. He therefore has to make out a case for the rational and truth-seeking character of such discourse. He does this partly by drawing up parallels between consensus truth with regard to norms, and consensus truth with regard to empirical science, where it may be held to be somewhat better understood.

In his 'postscript'[17] Habermas suggests a number of parallels between the knowledge claims of empirical and of hermeneutic science. Both derive from argumentation and the search for justifications and consensus in ideal communicative discourse, and each has its own 'domain of objects' to which such discourse refers. In the case of empirical science this is the domain of material objects in space–time; in hermeneutic science it is the domain of persons as participants in language, and their meanings, interpretations, evaluations, norms and goals. All these are objective contents of experience. Just as animals have objective perceptions, desires and satisfactions, so these become transformed by human language respectively into general statements, generalizable interests, and standards of evaluation. These universal domains of experience are objectivated (modelled) in the different forms of knowledge. There is, however, a difference in the way the object domains are appropriated by knowledge. Empirical knowledge starts with sense experience, issuing in observation and experiment (which is already discursive because descriptive), and then issues in theory building which is necessary for the intersubjective justification of empirical statements. Hermeneutic or communicative knowledge on the other hand has to start with the understanding of interpersonal relationships, which already includes the need to interpret linguistic utterances. These 'higher level' objects then become the object domain of hermeneutic knowledge. Both types of knowledge have a symmetrical relation to emancipation: empirical knowledge emancipates from the domination of a falsely understood nature, and communicative knowledge emancipates from social and personal distortions which are dominated by sectional interests.

There is a further point. In the past empirical scientific theories (such, for instance, as geocentric astronomy in the medieval and renaissance periods) have carried within them world views and evaluations which went far beyond empirical validation, and had clear ideological and social function. This property of theories has lost its claim to cognitive status in the pervasive positivism and instrumentalism of more recent interpretations. Habermas wonders whether a reunification of theoretical and practical (that is, communicative) knowledge might not result in part from a new cognitive interpretation of such world views:

> It has in no way been determined that the philosophical impulse to conceive of a demythologized unity of the world cannot also be retained through scientific argumentation. Science can certainly not take over the functions of world-views. But general theories (whether of social development or of nature) contradict consistent scientific thought less than its positivistic self-misunderstanding. Like the irrecoverably criticized world-views, such theoretical strategies also hold the promise of meaning: the overcoming of contingencies.[18]

Habermas' suggestion is that a view of theoretical science incorporating overall views of the natural world might be adopted, not as correspondences with the world, but as the media of human communication about nature in association with ethical and practical dimensions of world views. But it is a far cry from such a programme to the much more controversial claim that the same consensus theory of truth will restore cognitivity to *practical* discourse. Habermas takes for granted that in comprehensive historiography and social theory there are implicit 'world views', incorporating assertions about human nature and presuppositions about the good. An empiricist will have two objections to this account. First, it is not clear how the necessity of implicit world views follows from the fact that hermeneutic science has for its domain of objects persons and their mutual linguistic communications. For the fact that the object domain of a science contains persons' assertions of meanings etc. does not necessarily imply that the science itself is value-laden. It might be possible after all for investigators to give neutral 'spectator's' accounts and explanations of the meanings in the object domain. In order to appreciate the supposed connection between investigator's meanings and actor's meanings, it seems to be necessary to accept the standard argument about the irreducibility of person-talk to material-object talk, and to accept not only that person-talk is necessarily participatory and value-laden, but also that talk about person-talk is necessarily participatory and value-laden. Even if, with Habermas, we accept these assumptions (and I am inclined to accept them), the empiricist will still want to press a second and more difficult question 'Wherein lies the *objectivity* of these judgments of value?' Merely formal parallels with theoretical science in respect of universalizability and consensus in the ideal speech situation do not seem to be enough.

This appearance of insufficiency can be mitigated by one consideration deriving from Habermas himself, if I have understood his theory of truth correctly, and also by a certain reconstruction of his argument about the object domain of the hermeneutic sciences which is suggested by a somewhat different analysis of valuations in social science.

First, if the consequences of the consensus theory of truth for theoretical science are properly understood, it must be concluded that the 'objectivity' of *theories* is as badly off as the claimed objectivity of value judgments. For it is only the ghost of the correspondence theory that gives the illusion of firmer empirical foundations for theories than for evaluations. The conceptual frameworks of theories are *always* underdetermined by the empirical; if they are objective at all this is not where their objectivity lies. I have put forward elsewhere[19] severe objections against a form of realism which supposes that, even if our present theories are not true and no future theories will in fact be true, nevertheless it is meaningful to speak of an ideal theory that is in perfect correspondence with an external real world. The prop of the *noumenon* has been taken away with regard to theoretical science, and correspondence truth has been replaced by the ideal truth within theoretical discourse and empirical observation that has to be presupposed to permit scientists' own interpersonal communication about science to take place. In normative discourse similarly, the prop of an accessible realm of values has been taken away, and Habermas' suggestion is that it may be replaced by the ideal truth within moral discourse and argument as this has to be presupposed to permit practical agents to communicate about action.

Secondly, I have argued elsewhere[20] a somewhat different account of the relation between empirical and hermeneutic science which I think does better justice to their similarities and differences than that adopted by Habermas from most of the standard accounts. My thesis is that there is not so much a parallelism as a linear continuity between the empirical and the hermeneutic. They both have the same domain of objects, namely bodies, including persons' bodies, carrying their properties around in space and time. At each stage of the continuum, appropriate interpretive conditions enter the process of theorizing—formal and material regulative principles at all stages from physics onwards, then interpretations in terms of norms and deviances, stabilities and instabilities in biology, and finally evaluations incorporated in world views in the sciences of man and in history. The choice of 'persons' and participatory meanings as fundamental concepts in the hermeneutic sciences is not a necessary choice, as is shown by Habermas' own barely disguised fears that scientistic and impersonal 'systems theories' may after all prove technically successful in organizing post-capitalist society on a stable basis. The choice of the concept 'person' becomes 'transcendentally necessary' only *after* an option is taken for practical discursive rationality and individual humanity.

Habermas sharpens the dilemma in the context of a debate with systems theorists such as Luhmann, who effectively *deny* the assumption that communicative discourse about persons implies participatory value judgments. Systems theories are essentially scientistic accounts of human and social behaviour which supply the probabilistic and utilitarian grounds of social planning. Here the planners treat the planned as scientific objects, guided by value norms that are chosen on grounds extrinsic to the object domain, that is to say they are imposed by decisions, however moral in intent, that have nothing to do with participatory dialogue with those planned. Such norms are clearly in Habermas' terms not produced by consensus in ideal speech, and hence have no claim to 'objectivity'. Moreover it has been empirically observed, by Weber among others, that such planning will not be successful unless the systems of norms are internalized by those who are planned. The norms have to carry their own legitimation, and they are therefore developed into mythological world views or ideologies, and sold to the populace as necessary fictions. Or at least this cynical and self-conscious description of the proceedings is what is said to be required, now that its naïve and unself-conscious occurrence in earlier stages of society is no longer possible for us.

At this point Habermas introduces his view of religion. Religious views have traditionally provided a unified view of the social, human and natural cosmos which had two functions. First there is legitimation of authority structures necessary for social control, and second there is a personal universe of meaning which buttresses the individual against anomic forces and disasters, both natural and social. But for these functions to operate there must be individual *belief* in the world stories which perform the functions, so either they must be imposed as 'rationalizing illusions' (as Luhmann suggests),[21] or they must be founded on truth. The first alternative is contrary to the symmetries of ideal consensus and must be rejected. Religion on the other hand:

> by promising meaning . . . preserved the claim—until now constitutive for the socio-cultural form of life—that men ought not to be satisfied with fictions but only with 'truths' when they wish to know why something happens in the way it does, how it happens, and how what they do and ought to do can be justified.[22]

And one of the signs that humanity may evade the fate of the ahuman planned society of the systems theorists is to be found in certain developments in theology. There is a 'repoliticization of the biblical inheritance . . . which goes together with a leveling of this-world/other-worldly dichotomy', and although to Habermas this does not seem consistent with the idea of a *personal* God, nevertheless:

the idea of God is transformed [*aufgehoben*] into the concept of a *Logos* that determines the community of believers and the real life-context of a self-emancipating society. 'God' becomes the name for a communicative structure that forces men, on pain of a loss of their humanity, to go beyond their accidental, empirical nature to encounter one another *indirectly*, that is, across an objective something that they themselves are not.[23]

With regard to the objective character of such communication, the hermeneutic philosopher Gadamer comments that Habermas' theory needs more explicit metaphysical foundations, for there must be a metaphysical origin of the 'idea of right living' contained in the concept of unconstrained consensus.[24] Gadamer argues that hermeneutics as such is *not* subjective—there are real 'semantic structures' present to experience, and consensus of mutual understanding is more than conventional. Habermas agrees that to evade the subjectivity of hermeneutics it is necessary to refer to the social interest in interpretation and communication of practical affairs, but holds that Gadamer's approach is too traditionalist and conservative. Mutual interpretation of current norms and beliefs is not sufficient, since it provides no external point of reference from which to criticize both their and our own social ideologies. And to do this Habermas is forced beyond hermeneutics to talk of anticipations of the 'world as a whole', which can measure the claim of particular social orders in the setting of total history. Ideal discourse is anticipated in a context in which the search for freedom, justice, and truth go together, and hence commits the philosopher to certain forms of social praxis. As for Weber, so for Habermas, this commitment is to the inheritance of European liberalism and social democracy. At the end of his argument with the scientific systems theorists Habermas says:

Even if we could not know much more today than my argumentation sketch suggests—and that is little enough—this circumstance would not discourage critical attempts to expose the stress limits of advanced capitalism to conspicuous tests; and it would most certainly not paralyze the determination to take up the struggle against the stabilization of a nature-like social system [i.e. a scientistically controlled system] *over* the heads of its citizens, that is, at the price of—so be it!—Old European human dignity.[25]

Conclusion

In conclusion I have space to do no more than pick up one point which seems to me to pose a major question for Habermas' thesis. This concerns not so much the status of the ideal speech situation itself, as the reason for adopting it as the standard of non-empirical truth. There seem to be just five possibilities for answering the question 'What has ideal consensus to do with truth?'

1. It can be argued to be a viable theory of empirical truth which would permit a univocal understanding of cognitive terminology—such concepts as knowledge, truth, objectivity and the like become the same over the whole domain of the sciences. This is certainly a desirable characteristic of any theory of truth, but by itself it is not an argument.
2. Another possibility is that we are forced to the notion of truth as ideal consensus in interpersonal discourse, where we must treat the 'other' as a person having symmetrical rights with our own. This at best seems a negative argument rather than a positive interpretation of truth. That is, commitment to personal discourse does imply that we recognize *non*-ideal situations where constraints on argument occur, but gives no positive account of truth or of how we might be sure of recognizing it if it were attained. This is not to say that a theory of truth necessarily has to provide a practical recipe for its discovery, but if we are given no more than a counterfactual ideal, we have no means of judging whether the ideal is appropriately designated 'truth' in anything like the traditional senses. Why is ideal consensus a theory of *truth*?

Moreover, this argument seems to beg the question of why we must enter interpersonal discourse at all in the search for truth about norms. Truth *might* have been given to Moses on the mount, in which case it would be an offence against truth itself to offer symmetrical rights to all other men in a kind of democratic Quaker meeting. The assumption of 'persons' with equal rights is a part of the European liberal ideal which is itself correlative with the consensus theory of truth, and no empirical or transcendental arguments have yet been provided to show that we are *forced* to play this particular game with respect to all human beings. Positivists, systems theorists, and the like do not choose to do so with regard to the whole of humanity, and it can only plausibly be argued that they are 'forced' to do so in regard to their own fellow scientists when setting up their own scientistic theories. But in that case they are not forced to regard this discourse as constitutive of any truth other than empirical truth. For them norms may and do without contradiction remain arbitrary and groundless.

3. A further possibility is to regard the concept of ideal discourse as evolved from animal behaviour and human communication throughout history. Such an argument would purport to show that the concept is not an arbitrary option, but something which appears in its own good time in history. But like all ethical arguments from evolution this does not show that it is good or conducive to the truth that the concept *should* be acted upon. Moreover, any such interpretation of human evolution seems to fall foul of Habermas' own strictures against Chomsky's naturalism—it is itself a social theory and hence presupposes its own normative standpoint, which makes any argument *from* consensus *to* true norms circular.

4. Perhaps the definition of truth as ideal consensus is itself the result of ideal consensus. This suggestion can surely be briskly dismissed as at worst empirically false, and best counterfactual, and in any case patently circular.

5. There remains what seems to be the only viable possibility, which I am afraid Habermas would characterize as 'decisionism'. This is to adopt the ideal of discourse and its concomitant concept of 'person' as a moral standpoint, and to embrace the various forms of circularity described above as arguments which both support and are supported by that standpoint. The option for ideal discourse is not then ungrounded, but its grounds are not transcendentally conclusive. In this case, however, it seems that there might in principle be other such arguable options, and there certainly are other options in practice in our pluralist society. For if one moral standpoint can be adopted in terms of mutually

confirming and ultimately circular arguments, then others might be too. Habermas' attempt at a unified theory of truth is not sufficiently persuasive to rule out the possibility either of some other value-oriented unification of the claims of all the sciences (as for instance some Marxist theories of history are), or that, as positivists would hold, there is no truth apart from empirical science and logic. I would myself opt for the former possibility, though not in its Marxist version.

Notes

1 In the preparation of this essay I have been greatly helped by correspondence with Thomas McCarthy, and by his kindness in sending me the advance typescript of his book, *The Critical Theory of Jurgen Habermas*, Cambridge, Mass., 1978. Errors of interpretation that remain are entirely my own.

2 I am referring here to concern with the problems of theory-change and the theory-observation relation which have led to three types of response in recent philosophy of science:

(*a*) Pragmatic instrumentalism, in which theory-languages become games without truth value, and the essence of science is held to be pragmatic empirical success rather than descriptive truth.

(*b*) Concentration on ontology in place of epistemology, where truth is formally defined and either dogmatically asserted to inhere in a realistic interpretation of current science, or not accompanied by any epistemological decision procedures at all for determining the truth of actual theories.

(*c*) Relativism, which is the theoretical concomitant of (*a*), and in which no external empirical criteria of truth are recognized for theories, but truth becomes coherence within a given theory.

All these views presuppose traditional versions of either the correspondence or the coherence theory of truth, and objections to them become objections to these traditional theories. This sets the stage for the attempt to find in Habermas' work an alternative theory of truth in terms of consensus.

3 This debate is to be found in H. Marcuse, *One-Dimensional Man*, Boston, 1964, p. 166 and J. Habermas, *Toward a Rational Society*, London, 1971, p. 86.

4 Apart from *Toward a Rational Society*, cited above, primary sources for Habermas' account of natural science are his 'Towards a theory of communicative competence', *Inquiry*, vol. xiii, 1970, 360; *Knowledge and Human Interests*, London, 1972; 'A postscript to *Knowledge and Human Interests*', *Philosophy of the Social Sciences*, 1973, 360; *Theory and Practice*, London, 1974, chs 1 and 7; and 'Analytical theory of science and dialectics' and 'A positivistically bisected rationalism' in

The Positivist Dispute in German Sociology, ed. T. W. Adorno *et al.*, London, 1976, pp. 131, 198.
5 Habermas, *Knowledge and Human Interests*, p. 118.
6 Habermas, 'A postscript . . .' p. 176.
7 *Knowledge and Human Interests*, p. 4.
8 'Towards a theory of communicative competence' and 'A postscript . . .'
9 'Towards a theory . . .'
10 'A postscript . . .', p. 180.
11 T. McCarthy, 'A theory of communicative competence', *Philosophy of the Social Sciences*, vol. iii, 1973, 135.
12 Habermas, *Legitimation Crisis*, Boston, 1975, p. 108.
13 Quoted in McCarthy, 'A theory of communicative competence', p. 140.
14 Cf. Popper's account in his *Conjectures and Refutations*, London, 1963, pp. 228, 391, where it becomes clear that increasing 'verisimilitude' is not a property of identifiable sequences of theories in history.
15 Habermas, 'A review of Gadamer's *Truth and Method*' in *Understanding and Social Inquiry*, ed. F. R. Dallmeyr and T. A. McCarthy, Notre Dame, 1977, p. 335.
16 Luhmann, paraphrased in Habermas, *Legitimation Crisis*, p. 99.
17 Habermas, 'A postscript . . .', p. 172.
18 *Legitimation Crisis*, p. 121.
19 See M. Hesse, *The Structure of Scientific Inference*, London, 1974, chs 1 (reprinted as ch. 3 above) and 12, and *idem*, 'Truth and the growth of scientific knowledge', in *PSA 1976*, ed. F. Suppe and P. D. Asquith, vol. ii (reprinted as ch. 6 above). (Philosophy of Science Association, East Lansing, Mich., 1977).
20 See M. Hesse, 'Theory and value in the social sciences' in *Action and Interpretation*, ed. C. Hookway and P. Pettit, Cambridge, 1978, p. 1 (reprinted as ch. 8 above).
21 Habermas, *Legitimation Crisis*, p. 121.
22 *Ibid.*, p. 119.
23 *Ibid.*, p. 121.
24 Gadamer, *Hermeneutik und Ideologiekritik*, 1971, quoted in W. Pannenberg, *Theology and the Philosophy of Science*, London, 1976, p. 200.
25 *Legitimation Crisis*, p. 143.

IV SCIENCE AND RELIGION

10 *Criteria of Truth in Science and Theology*[1]

Faced with what he saw as the danger to society in the ascendancy of natural science and decline in religion and morals, the great French sociologist Emile Durkheim sought the origins of both religion and science in their function in primitive societies as guarantors of social solidarity.[2] In contrast to Frazer, Tylor, and other early anthropologists, he looked for the internal intelligibility of myth and ritual in social terms, rather than regarding them just as failed attempts to state objective truths about the natural world of the same kind as those later arrived at by natural science. One does not have to accept Durkheim's ultimately atheistic identification of God with Society, nor the politically authoritarian consequences which have sometimes been held to follow from this identification, to see that Durkheim has raised a set of issues of crucial importance for a discussion of science and the myths and doctrines of religion. For what he has done is to reintroduce a third factor into the perennial and by now rather frustrating dichotomy between science and religion, namely society and the sciences of man. It is some consequences of this approach that I want to explore in this chapter.

I shall begin by retracing some familiar ground: how the dichotomy between science and religion became fixed during the seventeenth-century scientific revolution; how natural science came to be seen as the epitome of objective knowledge, and consequently acquired a monopoly of the understanding of value-neutral truth, and how this has led to two extreme views to be found in current philosophical theology: either that theology is metaphysical and meaningless, or that its meaning and truth are defined internally in a certain sort of language-game which is one among many such possible games whose cognitive criteria are mutually independent. Then I shall consider some current critiques of the scientific notion of 'truth' presupposed in both these views, and ask how it is affected by the growing epistemological importance of the human sciences. By interpreting certain aspects of these sciences as value-laden ideologies, I shall consider how far Christian theology should also be understood as such an ideology, and, by the way of example, oppose its interpretation of the nature and destiny of man to those of two other influential modern systems to be found in the works of Jaques Monod and Louis Althusser.

I

The 'new way of philosophizing' that yielded the scientific revolution of the seventeenth century was the heir of two traditions which have remained in increasing though partly disguised tension ever since. These were on the one hand the tradition of rational cosmology inherited from Greek science, from which came the theoretical framework of the new science, and on the other hand the practical traditions of craftsmen, navigators, metallurgists and alchemists, from which came the concept of experimental test and the hope of technical control of nature. As Francis Bacon expressed it at the beginning of this scientific movement:

> . . . the matter in hand is no mere felicity of speculation, but the real business and fortunes of the human race, and all power of operation. . . . And so those twin objects, human Knowledge and human Power, do really meet in one; and it is from ignorance of causes that operation fails.[3]

From the beginning this dual inheritance exposed science to conflicting criteria. Cosmology, in its Aristotelian form, was until the Copernican revolution an expression of a world view. In the form in which it had been Christianized by Aquinas, it expressed a view of man as the crown of creation, playing out the drama of history and redemption on the stable stage of the earth-centred universe, round which revolved other globes, conferring upon the earth and upon man benefits of light, heat, and perhaps astral influence. The traditional alternative cosmology was the acentred universe of chance configurations of atoms moving arbitrarily in space, and this was generally rejected as chaotic, atheistic and morally degenerate. The accepted cosmology was therefore, in something like Durkheim's sense, part of an integrated system in terms of which man understood himself, his society and his history.

The emphasis on empirical test was bound to come into conflict with this function of cosmology. Empiricism introduced a notion of truth and knowledge in terms of which science was understood to have the aim of 'reflecting' an external world which was 'distanced' from man, and in principle wholly independent of him. The aim of science was to uncover 'the facts' irrespective of their consequences for man's value-system and view of himself and society. Theological and moral arguments for and against particular aspects of scientific theory began to be held to be irrelevant, although these arguments still played an important role in early seventeenth-century science, for example in the rejection of astrology, which was held to be in conflict with man's freedom as guaranteed by Christian revelation, and the rejection of teleological explanations in science, which were distinguished by Bacon as the province of theology, not science: final causes, he said, are like virgins dedicated to God, and therefore barren of empirical fruit for the good of man.[4] What has variously been called the 'disenchantment' of the world (Max Weber), and the 'dissociation of sensibility' (T. S. Eliot) became the presupposition of science. And not only of science, for the new concept of objective value-free scientific truth began to be transferred also to those aspects of theology that impinged upon the subject matter of science: cosmology, the history of the earth and of biological species, the nature of life, mind and soul, and the alleged providential activity of the divine in human affairs. Not surprisingly, since this concept had originated with and was tailor made to fit scientific knowledge, it caused difficulties when applied as a measure of theological assertion. The details of various tedious and ultimately self-stultifying conflicts between science and religion do not need rehearsing here. The upshot in modern philosophy of science and religion has been twofold: first the positivist reaction which rejects *all* assertion not subject to scientific test as 'meaningless', or at least not cognitively valid, not related to knowledge or truth; and second, the reaction of later Wittgensteinians, who postulate a multiplicity of 'truths', valid in different and independent language games, each integral to different 'forms of life', neither impinging on nor conflicting with each other, and between which no judgments of relative validity can be made. The first of these reactions is perhaps more bracing for theology, as involving open challenge to the right of theology to exist at all as a cognitive discipline. The second has encouraged an all too prevalent temptation towards a ghetto mentality among the religious, who are allowed in its terms to play their own games as long as their form of life does not intersect with that of the secular, scientific world.

II

Both philosophical responses have altogether lost touch with the social insights to which Durkheim directed attention. It is still something like bad form in philosophical contexts in the English-speaking world to draw the conclusion that the concepts of truth and knowledge which science began to exploit in the seventeenth century are the source of the trouble, and are not necessary and perennial concepts of truth and knowledge. However, at this point the pursuit of scientific and philosophical argument in its own terms has come to the rescue. For these very concepts of truth and knowledge are now under severe attack from many directions. One of the first sources of attack was the study of the anthropology of belief systems to which Durkheim himself contributed. At first, Frazer, Tylor, John Burnett on the Pre-Socratics, and many others, naturally judged these belief systems against scientific criteria, and found them to be simply false. But soon the interconnections of cosmology, myth, ritual, and social, economic and technical arrangements came to be seen as systematic parts of an integrated whole, for which external criteria of *scientific* truth and falsity were not directly relevant.[5] 'Truth within a belief system' came to be a richer concept than mere reflection of facts in empirically testable propositions. A similar understanding of truth has also become relevant in many other disciplines: the history of ideas, seen as the study of belief systems in 'rational' as well as so-called 'primitive' societies; the sociology of knowledge, as a study of the correlations of beliefs and social systems; and finally, within the paradigm of knowledge itself, the history of natural science, which has been interpreted by Thomas Kuhn[6] and others as a sequence of systems of internally related concepts, metaphysics and theories, in which the aspect of verification and falsification by empirical test is claimed to have only minimal effect upon the theoretical explanations that science from time to time puts forward as 'true'.

Since this last attack on traditional theories of truth has breached the citadel of empiricist knowledge itself, it has received rather more philosophical attention than the other examples I have given. Let me briefly outline its consequences for our understanding of the truth of science. The key lies in the seventeenth-century conflict between theory and empiricism. On the one hand the cosmological urge to describe the essence of nature yielded a multiplicity of highly sophisticated theories purporting to reveal hidden entities and processes as the real causes of all natural phenomena. On the other hand there was the desire that these theoretical explanations should increasingly approximate to universal truths which, unlike all previous non-empirical cosmological systems, would never be overthrown by subsequent developments. Tensions arose between these theoretical and empirical aspects of the new science, primarily for the single reason that no amount of empirical investigation can guarantee the truth of a theory, since this always goes beyond the limits of observable experience in size or time or place. And since the truth of theory can never be guaranteed, there is always a multiplicity of theories that will fit the facts more or less well, whose credentials will rise and fall with culturally accepted norms as well as with experimental developments, and which will each in all probability eventually be refuted and rejected from serious scientific consideration. No reflective scientist is likely to deny that the way science conceives the fundamental nature of things at any given time will be very different in subsequent science (that is to say, if science survives long enough), and that the further theories get from observable facts, the more they are *underdetermined* by the facts, leaving open a multiplicity of theoretical interpretations.

Carefully stated, I believe this conclusion is both inescapable, and also that it constitutes a definitive refutation of the seventeenth-century ideal of science, and hence of its concepts of empirical knowledge and truth. One apparent difficulty must first be dealt with. There is often resistance to accepting the conclusion on the grounds that science quite clearly is in some sense a progressive and not a constantly revolutionary enterprise. After all, it is held, science *is* a system of continuously accumulating knowledge. We have learnt a great deal about the world which we exploit in scientific technology. Knowledge is power; we have learned and therefore can control. Of course this is true. But far from entailing rejection of the conclusion regarding *theoretical* science, this progressive character of science rather shows us how the peculiar kind of knowledge attainable by science should be understood and distinguished from theoretical knowledge. It is the kind of knowledge precisely appropriate to prediction and control; it is what is learned by an organism which attempts to adapt to and change its environment by gathering data in experimental situations, processing the information gained, and learning to predict successfully further developments in the environment and the subsequent effects of its own actions. It does not yield truth about the essential nature of things, the significance of its own place in the universe, or how it should conduct its life.[7]

Two further questions immediately arise. First, if theoretical explanation is not determined by this instrumental aspect of science, what is its significance? And, secondly, must we now revive the notion of a non-empirical metaphysics in the attempt to answer the non-empirical questions, where metaphysics, and perhaps particular scientific theories too, are subject to their own criteria of truth different from the instrumental criteria of successful prediction and control? The questions are related, and I believe the key to both of them is to be found in the sociological rather than the purely metaphysical realm. Or rather, I believe that the Marxists and Durkheim are both right in seeing metaphysics itself as intrinsically related to social life, and not subject to independent, perennial, criteria of truth, although some Marxists may be wrong in seeing this relation as unidirectional from substructure to ideological superstructure. It is rather, as other Marxists will agree, a mutual and reciprocal interaction.

As an example of the view of science I have outlined, let us take the present state of physical cosmology, and consider whether it has any relation to a religious 'doctrine of creation'. In taking this example I am not of course presupposing that Christian doctrines of creation are concerned only with temporal origins and the large scale structure of the universe. Many other aspects of natural science are *prima facie* candidates for consideration—particularly those reductionist aspects of biology and psychology that seem to come into conflict with views of man as free agent in his individual, social, and historical life, in which he, and indeed the non-human world also, are held to be open to the personal, and therefore non-lawlike, activity of God. I shall return to some of these issues in section IV below.

Of all physical theories, cosmological theories of 'model universes' are most obviously highly underdetermined by the facts. Modern cosmology originates with Einstein's general theory of relativity, which is a theory of the deformation of space curvature by the distribution throughout the universe of mass-energy. The equations of this theory permit a solution predicting that the universe is expanding, and this prediction seems to be confirmed by the already observed 'red shift' of the stellar galaxies, which suggests that they are receding from us in all directions with a velocity increasing with their distance. Two consequences follow in this model with regard to time and space. First, when the distance becomes so great that the velocity of recession reaches the velocity of light galaxies become invisible to us, and hence there is a limit to the size of at least the observable universe. Second, extrapolating backwards in time, there was a time at which all matter was together in a very dense region of space, under physical conditions that we cannot imagine or hope to describe with certainty, and beyond which further extrapolation back in time is so speculative as to be practically unguided by any conceivable experimental evidence. The latest estimates for the date of this 'origin' of the universe as we know it is given in 'big bang' models as about ten thousand million years ago.

There are many alternative model universes in scientific cosmology, all more or less fitting the facts as we know them from our very limited observations in space and time, and between which further observational tests are sometimes possible but never anything like decisive. The best known rival to the 'big bang' theory has been the 'steady state' theory. This caused some mistaken theological excitement when it was first proposed, because it postulated a continuous creation of hydrogen atoms throughout space, thus replenishing the total matter-energy of the observable universe to balance the receding matter disappearing beyond the light horizon, and maintaining an overall homogeneity of distribution of matter throughout space and time without the singular origin point of indefinitely high density. But the latest evidence seems to count decisively against this model. There are also models involving periodic expansions and contractions of the universe, which may have the curious consequence that time goes backwards in phases of contraction relative to the forward time of our phase of expansion.

But though science fiction sometimes seems to have taken the place of religious myth in our society, neither serious cosmology nor science fiction is theology, and several important points need to be made about the significance of these cosmological models. First, as I have said, they are so severely underdetermined by anything we can observe, that it is inconceivable that we shall ever reach by normal scientific means a definitely true theory of the physical structure and history of the universe as a whole. Second, although sometimes particular models can be tested against observation and at least tentatively rejected as inconsistent, much the most important kind of criterion for the acceptability of such models is mathematical: models have to have a certain degree of mathematical simplicity to be manageable at all, and there are also other quasi-aesthetic criteria for the goodness of a mathematical model which have little to do with empirical fact.

Thirdly, criteria for the goodness of a model definitely do

not, in present-day cosmology, include anything that could be called theological. This is a fact about acceptable physics in our scientific society; doctrine does not influence cosmology, even in the sense of being a source of imaginative hypotheses. This of course has not always been so—for both Aristotle and Newton theology was a source of scientific hypotheses—and it may not always be so in the future, but if there is to be a change of scientific practice in this respect, there will have to be a radical change with regard to the admissibility of non-natural, or at any rate non-physical, entities. God was abandoned by Laplace as an astronomical hypothesis, not necessarily because Laplace was an atheist, but because introduction of a *deus ex machina* put a stop to testable science: such a God could produce any effects, and his presence or absence in particular effects could therefore not be verified or falsified by any empirical test. Moreover, theology would gain nothing by such reintroduction of the God-hypothesis, for it, like all speculations that suggest new scientific theories, would be subject to the fluctuations of scientific fashion, as for example the hypothesis of atomism has been.

The fourth point to notice is that there is in our society little influence of physical cosmology upon social ideology or mythology. The social and to some extent the biological sciences are now much more fruitful sources of meaning for man's understanding of his status in the world. The reasons for this are complex, dating back to the origins of biology and social science in the nineteenth century, and certainly including the new understanding of the significance of the physical sciences that I have been describing. Whatever these reasons may be, they have the consequence that attempts to derive support for or refutation of theological doctrines from physics will appear in our social climate to be as irrelevant as corresponding moves with respect to any other ideology would be. This is not to deny that there is quite profound science fiction that uses cosmology in the interests of theological, sociological, and psychological insights, as for instance in the films '2001', and the Russian 'Solaris'. But the point is that these insights come from quite other sources than physical science. What, then, are these sources?

So far I have suggested that sources of meaning are not to be found in the kind of metaphysical systems which were thought to share a univocal concept of truth with science. Science gives us a world set objectively over against us, indifferent to man's purposes, which, if we accept the discipline of detached experimental investigation, we can instrumentally control. There is indeed such a world (I do not wish to advocate any species of philosophical idealism), but instrumental control does not even penetrate to the scientific theories we need to manipulate the world, much less does it reveal to us its meaning. If we are to understand its meaning we must put ourselves, individually and corporately, back into the picture. But a word of warning here. In theological contexts what I have just said may appear neither startling nor controversial. We are accustomed to hearing that 'theological truth is not propositional', but is personal, having to do with commitment and trust. But it cannot be too strongly emphasized that *nothing* that has been said in the English-language theological literature has begun to come to grips with this problem of truth on a level of intellectual rigour that matches the English-language philosophical literature. I know this seems a harsh verdict, but whatever has been done in the German and French languages within *their* philosophical tradition, however accurate the translation of this into English, it is not a sufficient contribution to the English philosophical debate. Christian theologians of all people reject at their peril their incarnation within their own natural language and philosophical tradition.

In the nature of the case I am not going to be able to do more than scratch the surface of this problem. There are, I think, two directions in which we can look for aid. The first is in the sciences of man themselves, which find themselves in a very similar dilemma: their attempts to extend the methods of natural science into their own domain seem to be breaking down, and on the other hand it is very unclear what sort of theories about men and society should take the place of the instrumental objectivity of natural science. And second, and more obviously relevant to Christian theology, current crises in man's understanding of himself have thrust forward ideologies that clearly compete with the Christian understanding of man and his history, so that for the first time for several decades there is serious intellectual concern with theological questions that are being asked on levels that are ideological rather than scientific.

III

The view of the natural sciences that I have been presenting has two characteristics that are relevant in a comparison with the social sciences. First, the criterion of success in natural science is success in prediction and control. Second, fundamental theories or conceptual frameworks cannot be interpreted as attempts at realistic description of the hidden features of the external world, but rather as particular viewpoints upon the world, partially determined by interests other than predictive success. Not only do the theories themselves come and go, but so do the interests they serve: in one culture they may be theological and metaphysical, in others sociological, and in others (ours) the acceptability of many physical theories can be fully understood only in terms of internal aesthetic criteria. When the project was first conceived of extending the successful methods of natural sciences to the sciences of man, that is in the nineteenth century, it was inevitable that the truth criteria for social theory should be seen as identical with those for natural theory. In the then current *mis*understanding of the purpose of natural science, it followed that the search for social theory should become a search for objective, value-neutral, descriptive laws and explanations, in terms of which the primary subject of the social sciences, man, has to be seen as a purely natural phenomenon, identified solely in terms of observable products: behaviour, language, codified laws and customs, and the means and forces of economic production.

The project was a radical break with all previous theories of man and society, which had been heavily dependent on value-laden interpretations of man's place in the cosmos, and the relative significance of the individual and society. Also the project was unrealizable. Whether it was unrealized in practice because of the extreme complexity and rapidly changing character of the subject-matter, or whether it is unrealizable in principle because of fundamental differences between nature and man as subjects of investigation, is a dispute that has gone on rather inconclusively for a hundred years. But whatever may be thought of that debate, its terms of reference have now been shifted to a different level as a result of the new and clarified view of the natural sciences I have just outlined. For consider what happens if we apply the two characteristics of natural science as now understood to the social sciences on the assumption that the two kinds of science should be essentially the same. First, the aim of the social sciences becomes success in instrumental prediction and control. Second, the theories of the social sciences, which are certainly as underdetermined as those of the natural sciences if not more so, need determining criteria other than the instrumental, and these may come from the same theological, metaphysical and ideological sources as those of natural science throughout much of its history. Aesthetic mathematical criteria are unlikely to be relevant, since social theories are unlikely ever to be structured in as much mathematical detail as those of physics, but a new source of criteria should perhaps be added, namely the introspectively psychological —what Dilthey and Weber spoke of as *verstehen*, the understanding of human meanings as it were from the inside.

Recognition of these necessary features of social science theory carried us far from the notion of objectified descriptions of a dehumanized external world which in nineteenth-century empiricism were the goal of natural and social science alike. And consideration of the first characteristic of natural science, the criterion of successful prediction and control, raises questions in relation to the social sciences which have not been thought to arise for the natural sciences until very recently. *Ought* the aim of social science to be primarily prediction and control? Put like this, it becomes clear that an affirmative answer is itself a value judgment of very great importance for the project of a social science. Conflicts between social engineering and traditional political wisdom, between computerized systems-theories and theories that interpret man's place in society and history humanistically, immediately suggest themselves, and between these conflicting views of what a social science should be, value decisions will necessarily have to be made, and theological judgments, whether they are explicitly present or absent, will most certainly be relevant.

Perhaps fortunately for theology, this is becoming all the clearer because the attempt to produce value-neutral social science is increasingly being abandoned as at best unrealizable, and at worst self-deceptive, and is being replaced by social sciences based on explicit ideologies, or at least on explicit points of view related to particular interests in society.[8] Some of these have atheistic and dehumanizing consequences that are bound to conflict with a Christian understanding of man and society. I shall devote the rest of this paper to describing two of them, both deriving, as it happens, directly and indirectly from French Marxism. I do not wish to claim that these two examples are by any means typical of viewpoints taken up in social science theory, particularly in the English-speaking world, but to study some extreme cases may help better to elucidate the nature of the potential conflicts.

IV

My first example is the French molecular biologist Jaques Monod, and in particular his book *Chance and Necessity*.[9] Monod is an ex-Marxist, who has reverted from that largely non-scientific ideology to something much nearer to old-fashioned scientific humanism, but with a new and explicit recognition that this in itself involves non-scientific judgments of value and meaning. He attempts to show that modern evolutionary biology provides both an acceptable and comprehensive explanation of the origin and development of man and society, and that it is incompatible with any view of the universe as permeated by purposes relevant to human life, and to any view of man as the centre and purpose of the universe. Marxism as well as Christianity is an example of what he calls these 'animist' assumptions, since it, like Christianity, postulates the presence of purposes in history. Monod's basic premise is what he calls the 'principle of objectivity', that is, that only scientific knowledge counts as real knowledge, and he bases this on the level of instrumental control that science has made possible. Biological knowledge in particular, he claims, is now sufficiently established to make it certain that the development of living matter and its reproductive mechanisms, and the diversification of species by mutation and natural selection, are entirely governed by the operations of *chance* upon simpler primary material. Thus, although complex organisms like some higher mammals and man exhibit behaviour that can only be described in terms of their having purposes for their own life, this behaviour is entirely parasitic on the reproductive mechanisms that ensure the survival of invariant forms of the species in successive individuals. The development of these mechanisms is wholly explicable by chance, without postulating any suprahuman purpose immanent in nature. Biology has shown, he claims, that purpose is parasitic upon chance, whereas all 'animist' ideologies presuppose that purpose is primary, and even that it determines the forms of organic life. This refutation of animism, Monod goes on, explains the deep malaise of modern man, who is bereft of the value systems that used to seem objectively supported by suprahuman purposes. Man finds himself alone in the universe, the only source of his own value and purposes, just at the time when science has given him unprecedented powers to control his own environment and his history.

So much for Monod's argument. It culminates in a most moving and dignified presentation of man's dilemma, and Monod's conclusion, for which he explicitly does not claim scientific backing, is that the very practice of science itself demands moral decision, and is itself based on what he calls an 'ethic of objectivity'. This ethic is perhaps sufficient to replace the old ethics of animism: 'the ethic of knowledge that created the modern world is the only ethic compatible with it, the only one capable, once understood and accepted, of guiding its evolution' (p. 164). But can it calm the fear of solitude, satisfy the deep need, which is perhaps a genetic need for an explanation in which man is at home in the universe? Monod goes on:

I do not know. But it may not be altogether impossible. Perhaps even more than an 'explanation' which the ethic of knowledge cannot supply, man needs to rise above himself, to find transcendence. . . . No system of values can claim to constitute a true ethic unless it proposes an ideal transcending the individual self to the point even of justifying self-sacrifice if need be (pp. 164–5).

Many technical scientific and philosophical objections can be made to Monod's thesis which there is no room to develop here. What I want particularly to draw to attention, however, is first the level of argument upon which theological debate with Monod must be carried out, that is a level at least as scientifically informed and philosophically honest as his own, and second, that the debate is primarily and specifically one about values and action—how are we to live, given that the universe apparently answers to none of our aspirations and grounds none of our interpretations of ourselves and our significance? To reply to this thesis in terms of the theological categories of God, creation, providence, incarnation, and the rest, is not to debate about mere happenings or factual truths, or about theoretical speculations, but about meanings and values whose ultimate sanction, as Monod recognises very well, must be of the kind 'Here stand I, I can do no other', or perhaps 'Thus saith the Lord . . .'. But this does not mean a dogmatic reassertion of esoteric theological propositions, but a reaffirmation of insights which open out from where Monod and other secular men themselves actually stand.

If Monod's is an ideology of *chance*, my next example is an ideology of determinism. This is taken from the French hardline Marxist Louis Althusser. The writings of Althusser and his associates form a comprehensive philosophical corpus beside which the reflections of the scientist Monod are light indeed. It must be confessed, moreover, that modern French philosophical idiom is far removed from Cartesian transparency, and is particularly opaque to English eyes; it is therefore difficult to be sure that one has fully understood its implications. With that caveat, I shall try to give a brief summary of Althusser's explicit discussion and rejection of the Christian view of man.[10] I should first explain that Althusser uses the term 'ideology', as I have also been using it, to denote the theological or philosophical, political and social superstructure which he claims must accompany *every* economic system, including the socialist system. Thus 'ideology' is not in his view, as in many other Marxist and non-Marxist writers, a necessarily pejorative term denoting the *distorted* perspective of ideas that accompanies only imperfect social systems. It needs only to be carefully distinguished from what Althusser takes to be hard science, in a way similar to, but not identified with, the way I have distinguished the theoretical from the instrumental.

According to Althusser, the feature of Christian ideology that has persisted through the Enlightenment to modern secular philosophy and through feudalism to capitalist and post-capitalist economies, is the myth that God addresses men 'by name', as free individuals who in some sense initiate and are responsible for their own actions. In this myth God reveals himself to men as does the author of a book to his readers: the world is a book in which we can read God's nature and his will; the world is objectified and distanced from men precisely because it is God's creation into which men have been placed as independent agents. This is the real ground of the possibility of objective natural science as understood since the seventeenth century. In this sense the concept of 'man' presupposes the concept of God, and as Althusser's former pupil Michel Foucault develops the idea, the death of God is at the same time the end of man: it is 'the last man who announces that he has killed God.'[11]

So far Christians may find little to quarrel with in this interpretation. However, the next step is more radical. It is claimed that a necessary break has now occurred in the epistemological tradition, as a result of the development of the sciences themselves. The most characteristic feature of theories accepted in all the sciences is that they remove man from the centre of the universe. This was true of Copernican astronomy, in that man is no longer at the geographical centre; it is true of Marxian economics, where man is no longer the determining agent of his destiny; of Freudian psychology, where the ego is no longer the free centre of the self; and of modern linguistics, where even language makes man rather than man language. The consequent dehumanization of the world is very similar to the consequence drawn by Monod, but where Monod sounds a challenge to existential decision in this situation, Althusser and his colleagues ask only for 'correct' thinking reflecting 'correct' action, and both of these are determined by the structures of relatively transitory events as revealed by scientific theory, including Marxian theory. In this interpretation Marx is even deprived of his surrogate for divine providence: the inevitable march of history to some kind of social goal. There is instead only the passing scene with men caught up in the criss-cross of locally structured determinations.

Not only Christians but also secular philosophers and scientists have many resources for an intellectual critique of this position. I use it only as an example to raise questions that I cannot pursue here. But since it aspires to be a more or less coherent unity of theory and practice, two conditions must be binding on any critique. First, the position cannot be dismissed purely as crude and out of date materialist philosophy, because ideologies of theory and practice have a more complex hold on men than the purely intellectual. And secondly, ideology must be confronted with ideology. This will mean rethinking and restating Christian theology, not for the first time in its history, so as to confront modern heresy, not the heresy of the Greeks, nor of Islam, nor of the Enlightenment humanists.

V

Let me try to draw the threads together by summarizing what seem to me to be some concrete consequences for theology, and particularly for doctrines of creation.

The most important consequence is that the concept of 'truth' in theology, as also in much of social science, must be detached from the old empiricism of natural science. The assertions of theology make judgments of meaning and value which need to become explicit to prevent confusion with the concept of scientific truth. This is not so much in contradiction with the empiricist dualism of 'is' and 'ought', as a recognition that the instrumental criteria of science do not even guarantee the truth of scientific theory, let alone an account of God, man, and the universe.

There are two ways in which this conclusion may seem to

have disturbing consequences for theologians. First, since it is based partly on reinterpretation of the notion of theory in the social sciences, it may seem that it requires theology to be constrained within the social and ideological patterns of a given society. If this were a necessary consequence, then any view of theology leading to it should rightly be rejected. But it is not a necessary consequence, for the patterns of every society, certainly every advanced society, are complex overlappings of different influences and interpretations, some newly arising from new local social conditions, but some arising from history and tradition, and some from interactions with contemporary but culturally distinct societies. It is not that theology must reflect the implicit ideological judgments of our society, for there is no one consistent set of such judgments, but that it should arise out of and address itself to the real conditions of our society—it should be and be seen to be one of the *possible* expressions of our society, and not be, as so often now, a visible expression of archaism. This does not mean, of course, that it cannot depend at all on history or tradition or on the insight of other societies, for our society itself depends on these, perhaps far more profoundly than we may think.

Secondly, should the theologian be alarmed at the apparent rejection of the notion of perennial truths of which he has often been regarded as guardian? Again I do not think so. It may be that we shall have to abandon hope of expressing theological judgments in linguistic forms which, in the best translations, are claimed to be perennially valid. But this is not a surprising concession to have to make to cultural relativity: after all, language itself is heavily impregnated with changing ideologies, and even the language of physical theories is no longer claimed to be culturally invariant. It does not at all follow that perennial insights of the proper theological kind cannot be captured, however fleetingly and obscurely, in the appropriate expressions of each culture, any more than that it follows that the linguistic variance of physical theories prevents the facts that underlie instrumental control of nature having objective and perennial character. Care must be taken not to misunderstand this analogy—it is not intended to suggest a similarity in the notions of 'fact' in theological and instrumental contexts, but only that a subject-matter referred to does not have to have invariant linguistic expression in order to exist and have certain perennial characteristics. If there are certain perennial theological facts, then no alarm need be caused by their transitory expression in different societies, for it will presumably follow from the nature of these theological facts that every society will reveal aspects of them either positively or negatively, perhaps even through what may be called 'God-shaped vacua', when the society appears to be committed to no positive theological expressions at all. There is of course no guarantee that theologians always discern or succeed in expressing these facts in the forms of their society, and there are no conclusive tests of success here, but such uncertainty is part of the price men pay for anything that deserves to be called knowledge.

What I have said about the nature of theological assertion is but a programme for study by philosophers of religion—perhaps the most important intellectual task facing the Church. I believe that the constraints upon and criteria for 'goodness' of theological assertion lie primarily in the areas of meaning and value judgment, and not in the area of empirical fact. This has the negative consequence that theological assertions that have been thought to impinge on empirical fact and hence on natural science, such as doctrines of creation and providence, are very little related to natural science as now understood. Theologians should not, therefore, be over-concerned with the state of physical cosmology, or even with reductionist biology, except where these are being used illicitly (as sometimes, for example, by Monod) to support philosophical positions that must themselves be rejected from a Christian standpoint. There is also a positive aspect to the relation of modern physics and biology to ideology and theology, but it should be seen rather in the freeing of the imagination in the ways exploited by science fiction, than as constraining man's interpretations of himself in conformity with any one of the changing theories of natural science.

Some current challenges to Christian theology are also challenges to materialist and reductionist ideologies, and they go very deep. Can we, for example, retain the concept of a man as a free and partly non-natural agent in the absence of a concept of God? If not, are Christians committed to reintroduction of some explicit moral and theological categories into the sciences of man? We are at least committed to reconsideration of many sacred scientific cows such as the indefinite possibility of natural knowledge and control, and the moral and political neutrality of science. Again, what should be the practical (that is to say, moral and political) stance of such theological reinterpretation? Some of the points I have made may seem to align Christian theology too closely with a libertarian and individualist view of man—an ideology that has been associated with religiously unfashionable Western laissez-faire capitalism. But these are complex issues, in which there is no simple correlation between ideological positions and stereotyped right and left in the decisions of everyday politics. I leave these controversies open-ended, partly because it is not for me to pronounce on specifically theological consequences, but also because I believe we are in an essentially open-ended situation. We have not begun seriously to address ourselves to these questions in any deeply rethought theological context. To adapt Kant's aphorism, theology without practice may be empty, but practice without theology is blind, and at present we have too much of the latter. Doubtless we shall not all agree on answers to the questions I have posed and on many others. Fundamental theological disputes may well break out again, but this will be a sign of health, because it will show theology being again incarnated in its own time, and even perhaps differently in each of the many and deeply distinct social systems of our time. We shall not create the successors of Augustine and Aquinas overnight, but if we are faithful to our own concerns, in his own good time God may.

Notes

1 This paper was presented at the Conference on 'The Doctrine of Creation' held by the Society for the Study of Theology in Edinburgh, 8–11 April 1975. I am grateful for the comments of the participants in discussion on that occasion and also at the D Society in Cambridge, 9 May 1975.

2 Especially in *The Elementary Forms of the Religious Life*, English edn, London, 1915.
3 Francis Bacon, *Works*, ed. J. Spedding and R. L. Ellis, London, 1858, IV, 32.
4 *Ibid.*, p. 365.
5 For useful discussions see *Rationality*, ed. B. R. Wilson, Oxford, 1970 and *Modes of Thought*, ed. R. Horton and R. Finnegan, London, 1973. Philosophical presuppositions of the concept of 'truth within a system' have been developed in the work of W. v. O. Quine; see especially his *Ontological Relativity and other Essays*, New York, 1969.
6 T. Kuhn, *The Structure of Scientific Revolution*, 2nd edn, Chicago, 1970.
7 For a similar interpretation of natural science see J. Habermas *Knowledge and Human Interests*, English edn, London, 1972, especially chs 5 and 6 Appendix. I have attempted to develop in more detail the concept of scientific truth implied in this paper in my *Structure of Scientific Inference*, London, 1974.
8 For useful discussions of objectivity and value in the social sciences see *Readings in the Philosophy of the Social Sciences*, ed. M. Brodbeck, New York and London, 1968, part 2; R. Dahrendorf, *Homo Sociologicus*, London, 1973; A. MacIntyre, *Against the Self-images of the Age*, London, 1971; K. Mannheim, *Ideology and Utopia*, London, 1936; G. Myrdal, *Objectivity in Social Research*, London, 1970; and J. Rex, *Problems of Sociological Theory*, London, 1970.
9 J. Monod, *Chance and Necessity*, English edn, London, 1972.
10 As presented in L. Althusser and E. Balibar, *Reading Capital*, English edn, London, 1970.
11 M. Foucault, *The Order of Things*, English edn, London, 1970, p. 385.

Bibliography of work by Mary Hesse

* *denotes paper reprinted in this volume.*

1952

'Boole's philosophy of logic', *Annals of Science*, vol. viii, 61–81.

'Operational definition and analogy in physical theories', *BJPS*, vol. ii, 281–94.

1953

'Models in physics', *BJPS*, vol. iv, 198–214.

1954

Science and the Human Imagination: Aspects of the History and Logic of Physical Science, London, p. 171.

1955

'Action at a distance in classical physics', *Isis*, vol. xlvi, 337–53.

1958

'Theories, dictionaries and observation', *BJPS*, vol. ix, 12–28.

'A note on "Theories, dictionaries and observation"', *BJPS*, vol. ix, 128–9.

1959

'On defining analogy', *Proc. Arist. Soc.*, vol. lx, 79–100.

1960

'Gilbert and the historians', *BJPS*, vol. xi, 1–10 and 130–42.

1961

Forces and Fields: A Study of Action at a Distance in the History of Physics, London, p. 318.

1962

'History of physics', *American Oxford Encyclopedia*, vol. iii.

'Counterfactual conditionals', *Aristotelian Society Supplementary Volume*, vol. xxxvi, 201–14.

'Models and matter' in *Quanta and Reality*, ed. S. E. Toulmin, London, pp. 49–57.

'On what there is in physics', *BJPS*, vol. xiii, 234–44.

'History and philosophy of science in the early Natural Sciences Tripos', *Cambridge Review*, vol. lxxxiv, 140–5.

1963

Models and Analogies in Science, London.

Critical notice of E. Nagel: *The Structure of Science, Mind*, vol. lxxii, 429–41.

'Measurement in science', *History of Science*, vol. ii, 152–5.

'A new look at scientific explanation', *Rev. Met.*, vol. xvii, 98–108.

'Action at a distance' in *The Concept of Matter*, ed. E. McMullin, Notre Dame, Indiana, pp. 372–90.

Commentary on C. C. Gillespie: 'Intellectual factors in the background of analysis by probabilities' in *Scientific Change*, ed. A. C. Crombie, London, pp. 471–6.

'Analogy and confirmation theory', *Dialectica*, vol. xvii, 284–95.

1964

'Francis Bacon' in *A Critical History of Western Philosophy*, ed. D. J. O'Connor, London, pp. 141–52.

'Hooke's development of Bacon's method' in *Proceeds of the X International Congress of the History of Science*, Paris, pp. 265–8.

'Philosophical foundations of classical mechanics', Resource Letter, *Am. J. Physics*, vol. xxxii, (December).

'Induction and theory-structure', *Rev. Met.*, vol. xix, 109–22.

'Changing views of matter', *History of Science*, vol. iii, 79–84.

'Analogy and confirmation theory', *Philosophy of Science*, vol. xxxi, 319–27.

1965

Forces and Fields, reprint, Totowa, New Jersey.

*'The explanatory function of metaphor' in *Logic, Methodolgy and Philosophy of Science*, ed. Y. Bar-Hillel, Amsterdam, pp. 249–59.

'Statistical methods for inductive logic', *Cambridge Research*, vol. i, 27–30.

'Miracles and the laws of nature' in *Miracles*, ed. C. F. D. Moule, London, pp. 33–42.

'Aristotle's logic of analogy', *Philosophical Quarterly*, vol. xv, 328–40.

1966

'Hooke's philosophical algebra', *Isis*, vol. lvii, 67–83.

Models and Analogies in Science, reprint, Notre Dame, Indiana.

1967

'Galileo and the conflict of realism and empiricism' in *Atti dei Symposium Internazionale di Storia, Metologia, Logia e Filosophia della Scienza,* Florence, pp. 283–9.

'Action at a distance and field theory' in *The Encyclopedia of Philosophy*, ed. P. Edwards, New York, vol. i, pp. 9–15.

'Ether', *ibid.* vol. iii, pp. 66–9.

'Laws and theories', *ibid.* vol. iv, pp. 404–10.

'Models and analogy in science', *ibid.* vol. v, pp. 354–459.

'Simplicity', *ibid.* vol. vii, pp. 445–8.

'Void', *ibid.* vol. viii, pp. 217–8.

1968

'Fine's criteria for meaning change', *Journal of Philosophy*, vol. lxv, 46–52.

'Consilience of inductions' in *The Problem of Inductive Logic,* ed. I. Lakatos, Amsterdam, pp. 232–46 and 254–7.
'A self-correcting observation language' in *Logic, Methodology and Philosophy of Science,* ed. B. van Rootselaar and J. F. Stahl, Amsterdam, 297–309.

1969
'Confirmation of laws' in *Philosophy, Science, and Method: Essays in Honor of Ernest Nagel*, ed. S. Morgenbesser, P. Suppes and M. White, New York, pp. 74–91.
'Positivism and the logic of scientific theories', in *The Legacy of Logical Positivism for the Philosophy of Science,* ed. P. Achinstein and S. Barker, Baltimore, pp. 85–114.
'Ramifications of "Grue"', *BJPS*, vol. xx, 13–25.
'The encyclopedia of philosophy', Essay review, *BJPS*, vol. xx, 263–9.
'Talk of God', Essay review, *Philosophy*, vol. xliv, 343–9.

1970
'Theories and the transitivity of confirmation', *Philosophy of Science,* vol. xxxvii, 50–63.
'Duhem, Quine, and a new empiricism' in *Knowledge and Necessity,* Royal Institue of Philosophy Lectures, vol. iii, ed. G. Vesey, Harvester, Hassocks, pp. 191–209.
'Francis Bacon' in *Dictionary of Scientific Biography*, ed. C. C. Gillespie, vol. i, 372–7.
'An inductive logic of theories' in *Minnesota Studies in the Philosophy of Science*, vol. iv, ed. M. Radner and S. Winokur, Minneapolis, pp. 164–80.
'Is there an independent observation language?' in *The Nature and Function of Scientific Theories*, ed. R. G. Colodny, pp. 35–77. (Reprinted in this volume as 'Theory and observation').
'Hermeticism and historiography' in *Historical and Philosophical Perspectives of Science, Minnesota Studies in the Philosophy of Science,* vol. v, ed. R. H. Stuewer, pp. 134–60.

1971
'Whewell's consilience of inductions and predictions', *Monist*, vol. lx, 520–4.

1972
'Probability as the logic of science', *Proc. Arist. Soc.,* vol. lxxii, 257–72.

1973
*'In defence of objectivity' Annual Philosophical Lecture, *Proceeds of the British Academy*, vol. lviii, 275–92.
*'Reasons and evaluation in the history of science' in *Changing*

Perspectives in the History of Science, ed. M. Teich and R. M. Young, London, pp. 127–47.
'Logic of discovery in Maxwell's electromagnetic theory' in *Scientific Method in the Nineteenth Century*, ed. R. Giere and R. S. Westfall, Indiana, 86–114.
*'Models of theory-change' in *Logic, Methodology and Philosophy of Science*, ed. P. Suppes *et. al.*, Amsterdam, pp. 379–91.

1974
The Structure of Scientific Inference, London, p. 325.
Forze e Campi: il concetto di azioni a distanza nella storia della fisica, Milan.
'Method in Maxwell's electrodynamics', *Proceeds of the XIII International Congress of the History of Science*, Moscow, Section VI, 14–21.
'Worlds, selves, and theories', *Cambridge Review*, vol. xcv, 62–5.

1975
'Bayesian methods and the initial probabilities of theories' in *Minnesota Studies in the Philosophy of Science*, vol. vi, ed. G. Maxwell and R. M. Anderson, Minneapolis, pp. 50–105.
'Bayesianism and scientific inference', *Studies in the History and Philosophy of Science*, vol. v, 367–72.
'Lonergan and method in the natural sciences' in *Looking at Lonergan's Method*, ed. P. Corcoran, Dublin, pp. 59–72.
'Models versus paradigms in the natural sciences' in *The Use of Models in the Social Sciences*, ed. L. Collins, London, pp. 1–15.
'Models of method in the natural and social sciences', *Methodology and Science*, vol. viii, 163–78.
'On the alleged incompatibility between Christianity and science' in *Man and Nature*, ed. H. Montefiore, London, pp. 121–31.

1976
*'Criteria of truth in science and theology', *Religious Studies*, vol. xi, 385–400.
Discussion: 'Peacocke's "Reductionism"', *Zygon*, vol. xi, 335–7.

1977
*'Truth and the growth of scientific knowledge', *PSA 1976*, vol. ii, (Philosophy of Science Association) ed. F. Suppe and P. D. Asquith, 261–80.

1978
*'Theory and value in the social sciences' in *Action and Interpretation: Studies in the Philosophy of the Social Sciences,* ed. C. Hookway and P. Pettit, Cambridge, pp. 1–16.
Introduction to symposium on 'Internal and external causation of scientific ideas' in *Human Implications of Scientific Advance: Pro-*

ceeds of the XV International Congress of the History of Science, ed. E. G. Forbes, Edinburgh, pp. 59–63.

1979

'The ideological and theological debate about science: introduction and a statement of the issues', *Anticipations*, pp. 4–5 and 8–11, (January).

'What is the best way to assess evidential support for scientific theories?' In *Applications of Inductive Logic*, ed. L. J. Cohen and M. Hesse, Oxford.

*'Habermas' consensus theory of truth', *PSA 1978*, vol. ii, (Philosophy of Science Association), ed. P. D. Asquith and I. Hacking.

Commentary on C. F. von Weizsacker: 'The preconditions of experience and the unity of physics' in *Transcendental Arguments and Science*, ed. P. Bieri, R. P. Horstmann and L. Kruger, Dordrecht, pp. 159–70.

Stanton Lectures, University of Cambridge

These lectures on the philosophy of religion expand on points made at the end of the Introduction to this book.

1978 *Science and Religion: some epistemological problems*

1 Truth and the 'Alphabet of nature'.
2 Realists, relativists, and the growth of knowledge.
3 Emergence and evaluation.
4 Facts, values, and meanings.
5 Habermas' consensus theory of truth.
6 Miracles and the problem of providence.
7 Metaphor and the attributes of God.
8 Myth and truth.

1979 *Social Construction and Religious Reality*

1 The new science-religion debate.
2 Society is God.
3 Religious representations.
4 From function to meaning.
5 Totems and techniques.
6 Decoding symbolism.
7 The mediation of paradoxes.
8 Religious reality.

1980 *Science, Religion, and Symbolism*

In preparation.

Work in Progress

'A revised regularity view of scientific laws' forthcoming in *Science, Belief, and Behaviour: Essays in Honour of R. B. Braithwaite*, ed. D. H. Mellor, Cambridge, 1980.

'Science and objectivity' forthcoming in *Critical Essays on Habermas*, ed. D. Held and J. Thompson, London and Massachusetts.

Acknowledgements

'Reasons and Evaluations in the History of Science' was first published in *Changing Perspectives in the History of Science* ed. M. Teich and R. M. Young, Heinemann Educational Books Ltd, 1973. 'The Strong Thesis of Sociology of Science' was first published as part of the paper 'Models of Method in the Natural and Social Sciences' in *Methodology and Science* vol. 8, 1975. 'Theory and Observation' was first published as 'Is there an independent observation language?' in *The Nature and Function of Scientific Theories* ed. R. G. Colodny, Pittsburgh, 1970, and was reprinted in Mary Hesse's *The Structure of Scientific Inference*, Macmillan, 1974. 'The Explanatory Function of Metaphor' was first published in *Logic, Methodology and Philosophy of Science* ed. Y. Bar-Hillel, North-Holland Publishing Company, Amsterdam, 1965. 'Models of Theory-Change' was first published in *Logic, Methodology and Philosophy of Science* ed. P. Suppes et. al., North-Holland Publishing Company, Amsterdam, 1973. 'Truth and the Growth of Scientific Knowledge' was first published in *PSA 1976* (Philosophy of Science Association), vol. 2. 'In Defence of Objectivity' was first published in Proceeds of the British Academy vol. 58, 1973. 'Theory and Value in the Social Sciences' was first published in *Action and Interpretation: Studies in the Philosophy of the Social Sciences* ed. C. Hookway and P. Pettit, Cambridge University Press, 1978. 'Habermas' Consensus Theory of Truth' was first published in *PSA 1978*, vol. 2. 'Criteria of Truth in Science and Theology' was first published in *Religious Studies* vol. 11, Cambridge University Press, 1976.

Index

Index